CARTILAGE REPAIR AND REGENERATION

Edited by **Alessandro R. Zorzi**
and **João Batista de Miranda**

Cartilage Repair and Regeneration
http://dx.doi.org/10.5772/67903
Edited by Alessandro R. Zorzi and João Batista de Miranda

Contributors

Adrian J Cassar Gheiti, Neil G Burke, Theresa Michelle Cassar-Gheiti, Kevin J Mulhall, Mokhtar Mars, Gilberto Jaramillo-Rangel, Yareth Gopar-Cuevas, Alberto Niderhauser-García, Adriana Ancer-Arellano, Ivett Miranda-Maldonado, María-De-Lourdes Chávez-Briones, Laura Rodríguez-Flores, Marta Ortega-Martinez, Paul Saluan, Anthony Egger, Michael E. Hantes, Apostolos H. Fyllos, Lara Herrero, Penny A. Rudd, David Flanigan, Joshua Everhart, Nicholas Early, Flávio Alves, Renan Paraguassu De Sá Rodrigues, Andrezza Braga Soares Da Silva, Gerson Tavares Pessoa, Laecio Da Silva Moura Moura, Jacyara De Jesus Rosa Pereira Alves, Kássio Vieira Macedo, Robson Giglio, Francisco J. Blanco, Mohamed Khamis Tolba Mahmoud Abdalla, Sandra Barbalho, Marina Akuri, Mariana Barion, Elen Guiguer

Published by InTech
Janeza Trdine 9, 51000 Rijeka, Croatia

Publishing Process Manager Martina Usljebrka
Technical Editor SPi Global
Cover InTech Design team

First published February, 2018
Printed in Croatia
Legal deposit, Croatia: National and University Library in Zagreb

Additional hard copies can be obtained from orders@intechopen.com

Cartilage Repair and Regeneration, Edited by Alessandro R. Zorzi and João Batista de Miranda
p. cm.
Print ISBN 978-953-51-3788-7
Online ISBN 978-953-51-3789-4

For EU product safety concerns:
IN TECH d.o.o., Prolaz Marije Krucifikse Kozulić 3, 51000 Rijeka, Croatia,
info@intechopen.com or visit our website at intechopen.com.

Meet the editors

Alessandro R. Zorzi lives in São Paulo (Brazil), where he works as orthopedic surgeon and researcher at State University of Campinas (UNICAMP) and Hospital Israelita Albert Einstein. He obtained his MD degree at Ribeirão Preto Medicine School (University of São Paulo, USP) in 1999 and his MSc and PhD degrees at UNICAMP. He does research in the field of meniscus and cartilage regeneration and in the surgical treatment of osteoarthrosis. He has an interest in scientific divulgation and is the editor of the blog Femur-Distal. He is married to Giovana and is the father of Ana Laura and Davi.

Professor João Batista de Miranda is the chairman of the Division of Knee Surgery and Inflammatory Diseases, at the Department of Orthopedic Surgery, Campinas State University (Unicamp), Brazil. He performs teaching activities with medical students, orthopedic fellows, and postgraduating researchers. He also develops research and clinical care. He is currently the superintendent of the Unicamp Teaching Hospital. Prof. Miranda obtained the title of PhD with an experimental study on bone regeneration and on allografts. He has published several scientific articles in international journals and is the coeditor of the book *Bone Grafting* of InTechOpen.

Contents

Preface

This work is the result of a partnership that began in 2011, when I received for the first time the invitation to be the scientific editor of a book on bone grafting, by the still little publisher known as InTech. I remember very well the publisher's proposal to make the knowledge more accessible through the open access system. At that time, I decided to accept the invitation of the still young publisher, founded in 2004. The reason was the enthusiasm of Ms. Ana Pantar, editorial consultant, and Ms. Jana Sertic, publishing process manager, also, because I agreed to the need for a new and must fair publication system.

Now six years later, InTech has grown and thrived. This is the fourth book in which I am the scientific editor. I can say that my respect and warm approval for the quality of the publisher's work only increased.

In this book, entitled *Cartilage Repair and Regeneration,* I am pleased to work with a subject that has gained much notoriety. The hyaline cartilage is a tissue that challenges tissue engineering and regenerative medicine because of its avascular nature. The chondrocyte, the cell responsible for producing the extracellular matrix that confers the unique properties of the hyaline cartilage, is one of the most difficult cells to be cultured, as well as neurons and endothelial cells, because of their high degree of differentiation and specialization.

At the same time, the advancement of the life span of the population and the increase in the practice of sports activities have led to an increasing incidence of pains caused by problems associated with cartilage lesions.

In the eleven chapters of this book, the reader will find texts written by researchers working on advanced topics related to basic laboratory research, as well as excellent reviews on the clinical use of currently available therapies.

Alessandro R. Zorzi, MD, MSc, PhD and João Batista de Miranda, Prof
Department of Orthopedic Surgery
State University of Campinas, Brazil

Section 1

Basic Science

Chapter 1

Viruses: Friends and Foes

Penny A. Rudd and Lara J. Herrero

Additional information is available at the end of the chapter

http://dx.doi.org/10.5772/intechopen.71071

Abstract

In this chapter, we will review how viruses can be used to positively affect joints and cartilage of their hosts. Many viruses are arthrogenic, and cause persistent and debilitating arthritis. Even those viruses that are not typically arthrogenic can also cause bone lesions as secondary pathogenesis. Some of these foes include members of the alphaviruses, like chikungunya and Ross River viruses, the rubiviruses, such as rubella, and erythoparvoviruses, like parvovirus B19. Some more uncommon viruses, which can occasionally have detrimental effects on their hosts' joints, include herpes simplex virus, varicella zoster, mumps, human cytomegalovirus, avian orthoreovirus, and caprine arthritis-encephalitis virus. Despite some viruses having negative impacts on cartilage and joints, others have been used as an effective means of gene therapy for bone and cartilage repair. We will take an in-depth look at the current therapeutic strategies for treating arthritis using various viral vectors.

Keywords: viral gene therapy, cartilage and bone healing

1. Introduction: viral peptides/vectors used as gene therapy for joint repair

Viruses have long been used as vectors for gene therapy. Some of the more popular viral vectors include retroviruses, oncolytic viruses, lentiviruses, adenoviruses, and adeno-associated viruses to name just a few. They are used in a wide variety of fields and are able to treat a diverse range of diseases, including Parkinson's disease, many cancers, amyotrophic lateral sclerosis, genetic disorders, cardiovascular diseases, hemophilia, and central nervous system CNS diseases and disorders [1–5]. In recent years, there has been an increase in the development of viral vectors to treat the musculoskeletal system, including the joints [6].

Articular cartilage damage can result from a variety of insults, either from over usage, diseases and disorders, or accidents, and often leads to different types of arthritis including osteoarthritis (OA) [7]. Articular cartilage damage can cause swelling, pain, and subsequent loss of joint function. Due to its structure, cartilage does not usually regenerate after injury or disease, thus leading to loss of tissue and formation of a defect [8]. Cartilage is devoid of nerves, lymph, and blood supply, thereby explaining the limitations to self-repair. Current therapies targeted at treating articular damage have demonstrated variable results. These therapies include oral administration of a variety of components of the extracellular matrix, such as glucosamine or intra-articular injections of corticosteroids, biological agents (e.g., infliximab, etanercept), analgesics, and autologous blood products [9, 10]. Many approaches have also been investigated to help heal cartilage damage, including the use of viral peptides/vectors as a means of gene therapy for joint repair. These strategies mostly rely on overexpressing therapeutic factors or suppressing genes involved in joint destruction. In this chapter, we will examine the use of the severe acute respiratory syndrome (SARS)-coronaviruses (CoV), recombinant adeno-associated, and adenovirus vectors as well as retroviruses and lentiviral vectors for the treatment of joint repair.

2. SARS-coronavirus peptides

Coronaviruses (CoV) are potentially lethal viruses of the *Coronaviridae* family. They are positive-sense enveloped RNA viruses, which infect humans and animals. Two virulent strains, HCoV-229E and HCoV-OC43, were first identified in the 1960s from patients who presented with coryzal symptoms. Due to increased surveillance of CoV disease prevalence, other strains circulating in the population have recently been identified, including HCoV-NL63 and HCoV-HKU1 [11, 12]. Also, since 2003, more pathogenic strains of coronaviruses have been discovered, including severe acute respiratory syndrome (SARS) and Middle East Respiratory Syndrome (MERS), which predominantly infect the lower respiratory track and cause lethal pneumonia [13, 14].

Despite being pathogenic, coronaviruses have been used as both viral vaccine vectors and gene therapy vectors [15–18]. The uses of CoV as vectors range from delivering immunostimulatory cytokines and antigens to treatment of feline infectious peritonitis. A recent publication has shown the potential of using CoV vectors for the treatment of arthritis [19]. In this study, the authors demonstrated that the use of a small synthetic peptide (MG11, 11 amino acids in length) derived from SARS-CoV fusion protein was able to reduce inflammation in a collagen-induced arthritis (CIA) mouse model. Furthermore, MG11 also was shown to protect mice against bone and cartilage damage. A 14-day treatment regimen with a dose of 25 mg/kg was considered efficient at reducing arthritis in this common autoimmune animal model of rheumatoid arthritis. Histological analysis showed treated mice had no or very minimal inflammation and minimal cartilage damage in the joints of the paws. Knees and ankles also had limited inflammation, or no inflammation, and synovial membrane thickening did not differ from normal limits. The findings suggested that the decreased pathogenesis is due to the ability of MG11 to inhibit cytokine and growth factor production mediated by inflammatory T cells. This study is interesting and paves the way for potential usage of CoV peptides as a novel therapeutic to alleviate rheumatoid arthritis.

2.1. Recombinant adeno-associated virus vectors (rAAV)

Adeno-associated vectors (AAV) are frequently used as viral vectors for gene therapy. They are small nonpathogenic members of the *Parvoviridae* family and the genus *Dependovirus*. These members are nonenveloped viruses with a single-stranded DNA genome (≈ 4.7 kb) [20] and only about 20–25 nm in size [21]. They are safe to use as viral replication (the lytic stage) can only occur in the presence of a helper virus, either adenoviruses or herpesviruses. AAVs were first isolated from stocks of human and simian adenoviruses and thought to be contaminants [22]. AAVs are of interest since they have the ability to specifically integrate into host genomes and establish latent infections. Furthermore, more great advantages are that the preparations are stable and can be produced at titers of more than 10^{12} particles per ml [23].

Many clinical trials have commenced looking at the use of rAAV for treating a variety of conditions including but not limited to Pompe disease, cystic fibrosis, Parkinson's disease, muscular dystrophy, α-1 antitrypsin deficiency, and hemophilia [24–28]. Europe has even approved a rAAV drug manufactured under the name Glybera, which is the first gene therapy, to treat a very rare disease called lipoprotein lipase deficiency [29]. Despite the efficacy, the staggering cost of such a treatment has hindered the commercial success and use of this drug. There are currently a few other gene therapy drugs in the pipeline, including Amgen's FDA-approved drug IMLYGIC, which is a genetically modified oncolytic virus (Herpes simplex virus type 1) with proposed usage for melanoma cancers.

Previous work had shown that cell proliferation actually increases rAAV transduction, thereby making arthritis a candidate disease to be treated by rAAV [30]. Arthritis is not only accompanied by local influx of immune cells but also proliferation of cells in the synovial lining. The first *in vivo* experiment to examine the use of rAAV for the treatment of arthritis was done in the late 1990s. The authors chose to use a rat model of acute arthritis by intra-articular injection of lipopolysaccharide (LPS). The rAAV vector contained the *Escherichia coli* β-galactosidase gene regulated by the cytomegalovirus (CMV). The paper goes on to show efficient and stable gene delivery by rAAV and similar to previous *in vitro* findings, inflammation or disease state seems necessary to facilitate gene delivery. There is a clear enhancement of gene expression during the inflammatory process and the severity of the arthritis. At peak disease, 95% of the synoviocytes expressed high levels of the transgene, whereas when the arthritis subsided at 30 days post-LPS treatment, only basal levels of expression was seen. Interestingly, the study supports the feasibility of a preventative treatment approach, since rAAV responds to the disease state of the target tissue.

Since that study, rAAV vectors have demonstrated a great efficiency at transducing a variety of joint/articular cells, both *in vitro* and *in vivo,* including chondrocytes [31–33]. In the hopes of treating osteoarthritis, not only has the transduction of chondrocytes been investigated but also other important cells, including osteocytes, meniscal fibrochondrocytes, tendon/ligament cells, muscle cells, cells of the synovial lining and progenitor cells that may differentiate to form joint tissues [6]. The gene therapy approach is aimed at targeting a variety of mechanisms involved in the development of osteoarthritis, including cell proliferation and survival, the stimulation of anabolic pathways, the inhibition of inflammatory or catabolic pathways, and finally a combination of these strategies.

Approaches looking at stimulating growth and regeneration focus primarily on expressing known growth and cell survival factors, such as fibroblast growth factor-2, bone morphogenetic proteins (BMPs), telomerase, and antiapoptotic molecules like Bcl-2. Stimulating anabolic pathways involves building new molecules out of the products of catabolism. It is thought to aid in restoring function/production to the extracellular matrix (ECM), using growth and transcription factors or signaling molecules, for example, insulin-like growth factor I (IGF-I), parathyroid hormone-related peptide, Indian Hedgehog, SOX factors, etc. Whereas the inhibition of catabolic pathways uses inhibitors of matrix-degrading enzymes, inflammatory cytokines, as well as that of chondroprotective cytokines like IL-4 and IL-10.

Caution needs to be taken when trying to implement the use of rAAV vectors in humans as a large proportion of the population have antibodies against AAV, which would greatly hinder its therapeutic efficacy. However, most of these antibodies are against the serotype AAV2 [34]. With several different serotypes, often therapeutic strategies aim to engineer variants to generate vectors with improved tissue specificity and transduction efficiency, while also avoiding the effects of preexisting neutralizing antibodies [35].

2.1.1. Using rAAV to treat bone regeneration

Bone loss occurs in a wide spectrum of inflammatory diseases including rheumatoid arthritis (RA), coeliac disease, Crohn's disease, asthma, psoriatic arthritis, nephritis and myositis [36, 37]. Bone loss and associated sequelae greatly reduce the quality of life of many patients. Bone remodeling/regeneration is a dynamic and highly complex process involving a delicate interplay between osteoclasts and osteoblasts. Each year, our bodies regenerate about a quarter of trabecular and 3% of cortical bone [38].

Several studies have shown the ability of rAAV vectors to efficiently express bone morphogenic proteins into myoblast C2C12. Skeletal myoblasts, fibroblasts, and bone marrow-derived cells are pluripotent and can be stimulated with various BMPs (or other factors) to become osteoblast lineage cells [39–41]. These studies even showed relatively good success *in vivo*, where new bone formation was detected in rats between 3 and 8 weeks post injection [41]. More recently, rAAV was also examined to repair bone in a cranioplasty model [42]. Calvarial autografts and allografts were coated with 10^9 particles/mm^2 of rAAV2 vector expressing BMP-2 and transplanted into osteocalcin/luciferase (Oc/Luc) transgenic female mice. Microcomputed tomography (μCT) was used to measure the extent of bone formation, and findings showed that rAAV allografts resulted in significantly better bone repair. Furthermore, histological analysis also showed a variety of bone cells, as well as revitalization factors present in the grafts strengthening the conclusions of significant bone growth. However, the mechanisms involved in this AAV bone repair system still need to be elucidated.

Other studies have focused on expressing vascular endothelial growth factor (VEGF), receptor activator of nuclear factor κB ligand (RANKL), and constitutively active form of the activin receptor-like kinase-2 (caALK2) in rAAV vectors. Koefoed et al. also used AAV-coated allografts in a murine femur model [43]. This model is fairly popular where a mid-diaphyseal femoral segment is removed and replaced by an autograft, isograft, or allograft, which is secured by an intramedullary pin. In this report, authors used a frozen allograft that was coated on the cortical surface with 5×10^7 particles of rAAV, expressing caALK2. caALK2 can potently induce mesenchymal cell differentiation *in vitro* and *in vivo*, and its signals cannot be blocked by noggin or chordin, endogenous BMP antagonists. The results showed endochondral bone formation on the allograft. Interestingly, this procedure also prevented the formation of fibrotic tissue around the allograft, promoted blood vessel ingrowth, live bone marrow within the allograft, and stimulated osteoclastogenesis.

The group that opted to use rAAV expressing VEGF and RANKL did so because studies have shown that these factors significantly decrease during allograft healing [44]. Structural musculoskeletal grafts (i.e., bone, ligament), unlike other grafts, are often derived from allogenic cadavers. However, a significant drawback is that these transplants lack viability due to the absence of vascularization. This study aimed to examine that this rAAV could stimulate allograft vascularization and remodeling. The overarching hypothesis is that resorption of the graft through angiogenesis and osteoclast formation/activation leading to bone remodeling is a superior method to improve graft incorporation. VEGF/RANKL is known to regulate angiogenesis [45] and bone resorption [46] during skeletal repair. VEGF is secreted by hypertrophic chondrocytes and the perichondrium thereby recruiting endothelial cells and favor vascularization [47]. The data showed that if you block RANKL and VEGF signaling, there is indeed diminished bone formation on the autograft. A gain-of-function assay was also performed. RANKL and VEGF are sufficient to significantly improve healing by leading to a live, vascularized, remodeling.

Despite these positive results, more work is needed before this method can be used in humans. The connectivity between new and old bone needs to be ameliorated. In addition, technology allowing large animal, *in vivo,* 3D imaging of new bone formation and vascular ingrowth of allografts needs to be developed and biomechanical properties of rAAV-coated allografts must be determined and correlated with micro-CT parameters.

2.1.2. Using rAAV for cartilage repair

Cartilage is formed of connective tissue and found in many parts of the body, including joints. It is composed of chondrocytes surrounded by extracellular matrix, which contains glycoproteins, glycoaminoglycans, and structural and functional proteins. Articular cartilage is strong and flexible and protects the bones where they articulate to insure smooth movement and also absorbs shocks during weight-bearing activities. People with cartilage damage suffer from stiffness, pain, and swelling. Strategies for cartilage regeneration aim at modifying a variety of target cells including chondrocytes, synovial lining, osteocytes, meniscal fibrochondrocytes, tendon/ligament cells, muscle cells, and progenitor cells that may differentiate to form joint tissues [6]. Many rAAVs have been designed to target these cells.

Several papers have reported the ability to modulate cartilage both *in vitro* and *in vivo*. These studies aimed at over-expressing a variety of molecules like insulin-like growth factor-I (IGF-I), transforming growth factor-β (TGFβ), SOX-9, fibroblast growth factor-2 (FGF-2), antioxidant protein heme oxygenase-1 (HO-1), CTLA4-FasL fusion gene, bone morphogenetic protein-7 (BMP-7), dominant negative to Ikappaβ kinase β (IKKβdn), interleukin 38 (IL-38), interleukin-1-receptor antagonist (IL-1Ra), and osteoprotegerin (OPG) [32, 48–57]. These molecules can act on a plethora of functions, including enhancing cartilage anabolism (IGF-1, FGF-2, TGFβ, BMP-7), stimulating cartilage formation (SOX-9), exhibiting anti-inflammatory properties (CTLA4-FasL, HO-1, IKKβdn, IL-38, IL-1Ra), reducing oxidative stresses shown to exist is certain forms of arthritis (HO-1), and by blocking osteoclastogenesis (OPG). One paper examined using cystatin C (cysC) to inhibit cathepsin activity in the synovium of rabbit model of osteoarthritis. Unfortunately, this approach was unsuccessful. Despite completely blocking cathepsin activity in the synovium, synovitis, bone sclerosis and cartilage degradation remained [58].

Due to the large scope of these studies, for this review, we will summarize some of the main findings of rAAV and chondrocytes. The first attempts to transduce chondrocytes were in 2000. One group transduced primary human chondrocytes as well as human cartilage organ cultures with a rAAV-GFP. Their results were encouraging with GFP expression seen in more than 90% of monocultures after 7 days and over 45% of the cells in the organ cultures fluoresced for up to 28 days [59]. Around that same time, another group was looking at the ability of rAAV to be used *in vivo.* They used a rAAV-expressing bacterial beta-galactosidase (beta-gal) gene in an arthritis mouse model overexpressing tumor necrosis factor-alpha (hTNFalpha-Tg).

Another group also looking at transduction of a variety of primary human cells including tissues of mesenchymal, endodermal, neuroectodermal origin, and cartilage showed very different results. Chondrocytes appeared to have the lowest transduction rates along with dermal papilla follicles, epithelial cells, and fibroblasts. Transduction levels between 4.3 and 19.5% were seen [60]. Only melanocytes, G-CSF mobilized CD34+ and CD19+ cells fared worse, with no visible transduction seen.

Using genes, which are responsible for producing growth factors or molecules involved in cartilage repair, is a preferred method for viral therapy. Fibroblast growth factor-2 (FGF-2) is a member of the multifunctional fibroblast growth factor family and has broad mitogenic and angiogenic activities. One study examined whether rAAV is capable of delivering a functional FGF-2 gene cassette to isolated articular chondrocytes and to sites of articular cartilage damage *in vitro* and *in vivo* [61]. After encouraging results *in vitro*, the authors applied rAAV-hFGF-2 to osteochondral defects created in the patellar groove of knee joints in rabbits. Repair was seen at day 10 post infection and by day 20, the initial repair had progressed further, and integration into surrounding cartilage was seen. A follow-up at 4 months showed that the "new" cartilage now closely resembled the original tissue, but margins of new cartilage were barely visible. Results were even more encouraging as there were no apparent secondary effects such as synovitis or adverse reactions. Further histological analysis showed the absence of infiltrating cells at all time points observed. Earlier studies showed that rAAV could also transduce bone marrow-derived mesenchymal stem cells that migrate to injury sites [62]. Therefore, the mechanism of action is thought to be on two fronts: (1) rAAV can stimulate long-term FGF-2 transduction in damaged areas of cartilage, as well as in chondrocytes found in surrounding healthy areas and (2) rAAV also transduces the mesenchymal stem cells that will be recruited to damage areas and commence tissue repair. In 2008, *in vitro* studies using a combination of FGF-2 and SOX-9, a transcription factor that activates the expression of major cartilage matrix components, were also undertaken [53]. The premise behind this is that due to the complex nature of osteoarthritis and the plethora of processes involved in this pathology, efficient cartilage repair may require expression of several therapeutic factors. Toward this, 3D cultures and cartilage explants were used. rAAV-FGF-2 showed greater transduction efficiency and effective expression of FGF-2. While the addition of SOX-9 was equally efficient, it did not add to the overall effectiveness of the expression. The authors did not test but did suggest that repeated administration of the combination might improve the outcomes of cartilage repair over time.

The most recent report in the literature examined the use of polymer micelles in aiding rAAV as gene therapy. The polymer micelles enhanced the stability and bioactivity of rAAV, leading to higher levels of transgene expression in human OA chondrocytes *in vitro*. It was also found to aid in human osteochondral defect cultures to mimic a more natural environment. In addition, the micelles protected the viral vector against neutralization of the viral capsid. No detrimental effect on cell viability was observed when delivering rAAV/micelles to the cells at any time point of the analysis.

An investigation looking at the use of adjuvants for *in vivo* rAAV articular cartilage gene therapy has also been done. One group showed that light-activated gene transduction (LAGT) could be one such method. UV light accelerates the formation of the double-stranded transducing rAAV vector episome by activation of a host DNA polymerase. The use of UV exposure at doses of up to 200 J/m^2 actively increases transduction efficiency and expression of the transduced gene eGFP in cultures of immortalized and primary human articular chondrocytes, as well as articular cartilage explants. Importantly, this amount of light was noted as insufficient to cause harm to cells [63]. A follow-up study looking at the ability of UV light-activated gene transduction (LAGT) in chondrocytes *in vivo* showed that in rabbit chondrocyte cultures, as well as in intra-articular transduction of rabbit knees, LAGT treatment resulted in higher efficiencies compared to nonirradiated samples [64]. However, after 3 weeks, the mean fluorescence intensity of positive cells of the non-LAGT group had increased to the same level as the LAGT group, despite the proportion of transgene-expressing chondrocytes were still higher in the LAGT group. Overall results showed that LAGT probably does not benefit healthy cartilage. However, in diseased tissue, more chondrocytes were transduced in general and especially those close to the irradiated surface respond to the treatment. Importantly, further investigation needs to be done to assess if the biological effect is sufficient to provide a desired metabolic response toward repair.

Despite being a promising avenue for gene therapy, consistency among findings and systems using rAAV appears to be difficult and unpredictable. Further experimentation and stringent conditions will need to be done to establish if this treatment strategy is a viable and promising avenue to promote cartilage restoration.

2.2. Recombinant adenovirus vectors (rAdV)

Adenoviruses (AdV) are medium-sized nonenveloped viruses, composed of a nucleocapsid and a double-stranded, linear DNA genome of approximately 36 kb. Over 50 different human serotypes can be found and they cause 5–10% of all childhood upper respiratory infections. Adults can also suffer from illness caused by adenoviruses, but disease is generally mild and resembles that of a common cold.

AdV are interesting because they can infect a broad range of human cells and tend to yield high levels of gene transfer compared to levels achieved with other currently available vectors. This also includes high *in vitro* gene transfer efficiencies in chondrocytes and mesenchymal stem cells [65, 66]. These viruses can accommodate large genomic insertions up to 14 kb and have the ability to transduce these genes in both proliferating and quiescent cells. At least three regions of the viral genome can accept insertions or substitutions of DNA to generate therapeutic vectors. Also, the viral genome is relatively stable and undergoes limited rearrangements and inserted foreign genes are very well maintained through successive rounds of viral replication. Genetic manipulation of these vectors is easy by using standard recombinant DNA techniques, and they are easily grown, reaching titers of up to high up to 10^{13} particles/ml. Taken together, these factors make adenoviruses excellent candidates for viral gene therapy.

Human serotypes AdV2 and AdV5 from group C are the classic adenoviruses used as therapeutic vectors. Early versions of adenovirus vectors were unsuccessful due to the deletion of E1 region to accommodate the therapeutic transgene and to prohibit viral replication [67]. This deletion led to a strong innate immune response followed by adaptive responses, which destroyed the transduced cells, thereby defeating the purpose of gene therapy. Second generation vectors were generated by deleting several areas of the genome and allowing a larger amount of DNA to be inserted. Unfortunately, these vectors still triggered immunogenicity and led to cell death. Third generation vectors were known as "gutted" vectors. All viral coding regions were removed to prevent an immunological trigger. However, they need a helper vector that codes for the viral genome to allow for replication. These third-generation vectors facilitate insertion of up to 35 kb of genetic material and are therefore deemed high capacity. Gutless AdV have been delivered to different tissues in rodents, dogs, and nonhuman primates. These third-generation vectors have been shown to be nonimmunogenic for the life of a mouse, whereas the first generation induced a response within 3 months [68].

Along with rAAV, AdV is a very popular choice for gene therapy delivery. Much work has been done in a variety of fields including cancer, metabolic diseases, motoneuronal injuries/diseases, and cerebrovascular diseases. One of the first reports in the early 1990s examined the ability of AdV vectors to be useful tools in overexpressing anti-inflammatory molecules in rabbit synoviocytes to alleviate rheumatoid arthritis. Synoviocytes were chosen due to the ease of access via the intra-articular space and their longevity (type A, macrophage-like synoviocytes are estimated to live for 3–6 months) making them ideal candidates for viral transduction [69]. This study showed the ability of rAdV vectors to express lacZ via different techniques including *in situ* staining, immunohistochemistry, and transmission electron microscopy. The transduction remained detectable for over 8 weeks; however, efficiency did wane over time. Clinically, the rabbits fared well with no signs of arthritis, synovitis, or adverse effects for up to 8 weeks post-transduction, despite having preexisting antibodies to either human or rabbit adenoviruses. The authors were unable to identify exactly to which one the animals were previously exposed to, human or rabbit adenoviruses, since antibodies against rAdV are crossreactive against many species including humans, rabbits, and cattle.

A follow-up study by the same group looked at replacing lacZ expression with that of human interleukin-1 receptor antagonist protein (IL-1ra) [70]. IL-1 is an important mediator of inflammation and plays an important role in the pathogenesis of rheumatoid arthritis. IL-1ra is a natural receptor antagonist that competes with IL-1 for binding to type I IL-1 receptors and as a result blocks the effects of IL-1 [71]. Again, authors used New Zealand white rabbits as an *in vivo* model. After verifying *in vitro* that the expression of IL-1ra is biologically active, they found that direct intra-articular injection of rAdV into the synovium of rabbits led to the expression of high levels of IL-1ra within 1 week, as determined by Southern blot. However, like in their previous work, within 4 weeks, the levels of IL-1ra expression within synoviocytes decreased a major limitation to the approach. However, it is noteworthy to mention that these studies were undertaken using first generation vectors, which as mentioned above, are associated with major drawbacks.

Since the first studies in the 1990s, a plethora of publications regarding AdV have been published. Similar to rAAV, the focus of these studies seems to be targeted mainly on either bone or cartilage repair. Many studies have been interested in using rAdV to transduce bone morphogenic proteins including BMP-2, BMP-4, BMP-7, and BMP-13. In addition, soluble growth factors like PDGF, FGF and IGF, anabolic factors like growth factor and PTH, systemic angiogenic factors like VEGF as well as transcription factors associated with bone- and cartilage-related gene expression like Runx2, SOX9, osterix, and extracellular matrix molecules associated with induction or repression of mineralization like Gla protein, osteopontin, and bone sialoprotein.

2.2.1. *Using rAdV to treat bone regeneration*

Most of the studies using rAdV focus on the transduction of the various BMPs. One study by an American group based in Chicago investigated the feasibility of using a recombinant AdV to express 14 different bone morphogenic proteins (BMPs) [72]. It is known that bone demineralization can induce *de novo* synthesis of bone formation [73]. BMPs have been demonstrated to be the factors involved in bone regeneration. They belong to the TGFβ superfamily and are important in embryogenesis as well as in bone modeling. There are at least 15 different BMPs in humans, and this study attempted to establish which BMPs were the most effective at bone regeneration. The authors first examined the ability of the rAdVs to express ALP (an osteogenic marker) in the C2C12 cell line that is a precursor of osteoblasts. Four days after transduction, five BMPs were able to express ALP. These were BMP-2, BMP-4 BMP-6, BMP-7, and BMP-9. Findings were similar when looking *in vivo* at athymic nude mice. AdV were used to transduce C2C12 *in vitro,* and cells were then injected into the quadriceps muscle. Ossification was seen in animals that received AdBMP-2, 6, 7, and 9. However, BMP-7 was less robust than the other BMPs, and interestingly, BMP-6 and 9 were the most efficient. Since this study, numerous others have investigated the use of these bone-regenerating BMPs for *in vitro, in vivo,* and clinical studies [74].

Two studies of interest showed the ability of AdBMP-7 and AdBMP-2 to form bone intramuscularly and subdermally in immunocompetent rodents. A key factor in the success of these studies was to reduce immune responses to the adenoviral vector. Strong immune responses can decrease or inhibit therapeutic transgene expression. It was found that when the vector is delivered in conjunction with a collagen carrier, the vector becomes more effective in decreasing immunogenicity [75, 76]. Another method of prolonging transgene expression is by administering anti-T cell receptor monoclonal antibody following adenovirus-mediated *in vivo* gene transfer [77].

One of the most recent publications examined the effect of AdBMP-2 on the osteogenic ability of human mesenchymal stem cells (hMSCs) [78]. MSCs are multipotent somatic stem cells that are able to differentiate into a variety of cell types, including chondrocytes, myocytes, osteoblasts, and adipocytes. Targeting these cells with BMP-2 could potentially lead to their osteogenic differentiation and promote bone healing. *In vitro* experiments showed that when treated with AdBMP-2, hMSCs change phenotypes and resemble osteoblast-like cells. Further analysis showed that these changed cells also expressed ALP, an enzyme present in osteoblasts and critical for bone mineralization and calcification. Immunohistochemistry using a von Kossa stain (used for the quantification of mineralization in cell culture and tissue sections) showed increased positive staining at d14 post-treatment. Taken together, this study showed the potential of AdBMP-2 to skew the differentiation of hMSCs toward osteoblast-like cells, thereby potentially becoming a novel treatment for delayed or nonunion fractures.

In addition to BMPs, several other factors have been investigated to determine if their expression via an adenoviral vector leads to bone healing. Nell-1 is a novel direct transcriptional target of runt homology domain transcription factor-2 (Runx2). Nel-like molecule-1 (Nell-1) is osteoinductive on cells of the osteochondral lineage. Adenovirus vectors containing Nell-1 was shown to promote osteoblastic differentiation in calvarial cells (from the skull cap) [79]. An *in vivo* study demonstrated that Null-1 could be as efficient as BMP-2, one of the most potent BMPs, to induce rat calvarial bone formation [80]. VEGF, Sox9, Core binding factor alpha 1 (Cbfa1), Runx2, and noggin have all been investigated with varying degrees of success.

2.2.2. *Using rAdV to treat cartilage regeneration*

A recent publication showed that AdBMP-2 stimulates chondrogenesis of equine synovial membrane-derived progenitor cells. Chondrogenesis was determined by the up-regulation of collagen II, X and aggrecan, as well as the secretion of sulfated glycosaminoglycans and production of alkaline phosphatase [81]. Two other growth factors, Insulin-like growth factor-I (IGF1) and human growth and differentiation factor-5 (GDF-5), have also been examined for cartilage regeneration using rAdVs. IGF-1 is the major anabolic mediator for articular cartilage and plays an important role in maintaining cartilage homeostasis. IGF-1 enhances cartilage matrix metabolism by increasing the production of aggrecan, hyaluronan, and proteoglycan link protein-1 and by preventing degradation of proteoglycans. It also protects cartilage from the harmful effects of interleukin-1 or TNF following assault or injury. In one study, the ability of adenovirus vector encoding equine IGF-1 (AdIGF-1) to heal cartilage in an equine femoropatellar joint model was examined [82]. Then, 2×10^7 AdIGF-1-modified chondrocytes were injected into the joint and the animals were monitored for repair over the course of 8 months. The results showed that the AdIGF-1-modified chondrocytes were able to induce high levels of IGF-1, which persists for up to 9 weeks post-transplant. The increase in IGF-1 also led to an increase in collagen II expression. Histological analysis of tissue repair showed significant amelioration over control joints. Furthermore, no difference in inflammation was seen between naive chondrocyte-implanted or AdIGF-1-transduced repair tissues. These data were determined by examining inflammatory markers (including MMP-1, MMP-3, MMP-13, and aggrecanase-1) by qPCR. In addition, it was shown that IGF-1-enhanced repair also involved an increase in tissue thickness. It appears that there was a greater defect filling, and upon examination, these cells morphologically resembled chondrocytes rather than a fibro-cartilaginous-like phenotype seen within the control tissues. Another study in humans looked at the effects of AdV gene transduction FGF-2, FGF-2 combined with interleukin-1 receptor antagonist protein (IL-Ra), and/or insulin-like growth factor-1 (IGF-1). This was determined in both human osteoarthritis (OA) chondrocytes as well as in a leporine OA model [83]. FGF-2 expression protected human OA chondrocytes and decreased cartilage degradation *in vivo* (rabbit model). *In vitro,* FGF-2 induced collagen type II and an increased production of GAG. Furthermore, combining all three factors FGF-2, IL-1Ra, and IGF-1 leads to significantly lower levels of ADAMTS-5, MMP-13, and MMP-3, and increased amounts of TIMP-1. This was also true as seen in the rabbit model. The combined therapy seems to have a synergistic effect to achieve optimal results. The trigene expression system appears to promote GAG synthesis of chondrocyte, increases TIMP-1 expression, and reduces ADAMTS-5, MMP-13 and aggrecanase expression. Haupt et al. also found that an adenovirus-mediated gene therapy combining several factors was more efficient. In this study, IGF-1 and IL-1Ra were shown to promote the healing of cartilage injury in degenerative joint diseases, suggesting combination therapy could be beneficial for cartilage repair in degenerative joint diseases [84].

GDF-5 has been shown to be essential for normal appendicular skeletal and joint development in humans and mice. It positively regulates differentiation of chondrogenic tissue through its binding with bone morphogenetic protein receptor type 1 A and B (BMPR1A and BMPR1B). It also negatively regulates chondrogenic differentiation through its interaction with noggin (NOG). One study conducted by Luo et al. investigated the effects of adenovirus-mediated GDF-5 (AdGDF-5) on ECM expression in human degenerative disc nucleus pulposus (NP) cells in order to determine if AdGDF-5 is a viable therapy to treat intervertebral disc degeneration (IDD) [85]. Like many other studies, they began by determining the expression of GDF-5 *in vitro* after treating HEK293 cells with AdGDF-5 and then determined the optimal amount of viral vector needed for efficient transduction of NP cells. Following this, they investigated the effects of expression of GDF-5 had on the ECM. It was noted that GDF-5 promotes the synthesis of sulfated glycoaminoglycans and hydroxyproline, two major structures forming the ECM network. In addition, immunohistochemistry showed an increase in proteoglycans in the AdGDF-5-treated NP cells, stimulated NP proliferation, and increased the expression of collagen II and aggrecan genes. The outcome of this study indicates that NP cells within degraded discs would be ideal targets for the transduction of transgenic proteins and that AdV therapy could be a promising new avenue for the treatment of disc degeneration.

As like for rAAV, the effects of Sox9 on MSCs have been examined as a novel treatment of cartilage repair. This is of no surprise, considering that Sox9 is considered a master regulator of chondrocyte phenotype [86]. Like that of rAAV, Sox9 has been shown to be able to modulate cartilage both *in vitro* and *in vivo*. In the study led by Cao et al., Sox9 expression successfully promoted a chondrocyte morphology after AdV transduction of rabbit bone marrow mesenchymal stem cells (BMSCs) [87]. Overexpression of Sox9 resulted in the upregulation of collagen II and aggrecan, while inhibiting osteogenic differentiation. The latter was shown by a decrease in ALP staining and reduced expression of Runx2, Col I, and osteopontin. In rabbits, the AdVSox9 group had a better outcome regarding cartilage repair. This was seen by integration of *de novo* cartilage tissue repair, cells in the repaired tissue had distinctive morphology resembling chondrocytes that were surrounded by matrix that stained positive for safranin O and type II collagen. Finally, overexpression of Sox9 led to suppressing makers of hypertrophic chondrocytes (ColX and osteocalcin), thereby avoiding cartilage calcification.

In summary, like for rAAVs, rAdVs show a promising future for gene therapy to treat, or limit, joint damage. They have the advantage of growing to high titers, allowing high transduction efficiencies in a variety of cells and have shown promise in animal experiments as well as in explants. However, the main drawbacks for AdVs remain a long-term efficiency and overall safety. Prior exposure to various strains results in robust host immune responses against the vectors, greatly hindering long-term transgene expression in targeted patients. Moreover, the first patient death associated with gene therapy trials was that of an 18-year-old boy receiving a rAdV [88]. This vector contained ornithine transcarbamoylase (OTC), an enzyme needed to eliminate ammonia, and essential to treat the patient's partial OTC deficiency, which was present since birth. Unfortunately, the boy died 4 days after receiving the infusion and this adverse effect sparked controversy and ended in a lawsuit and formal investigation. Despite being the only death in nearly 4000 gene-therapy patients (over 400 trials), this hindered progress and saw extra measures for monitoring, reporting, and obtaining informed consent. The FDA and participants will probably still err on the side of caution when it comes to these types of clinical trials.

2.3. Retroviruses and lentiviral vectors

Lentiviral vectors are members of the *Retroviridae* family. These vectors can deliver a substantial amount of genetic information by spontaneously penetrating the intact nuclear membrane and inserting the "carried" DNA into the host's DNA. Due to this unique property, they are among the most efficient methods for gene delivery. Furthermore, they can integrate into either actively replicating or quiescent cells. For these reasons, they are commonly used for *in vivo* delivery of genome editing therapies. However, this ability to integrate into the host's DNA also raises a number of safety and ethical concerns. Another drawback of this class of vector is the possibility to activate tumor genes and to provoke insertional mutagenesis events upon integration. Examples of most frequently used lentiviruses include human, simian, and feline immunodeficiency viruses (HIV, SIV, and FIV).

Compared to other forms of viral gene therapy, the main advantages of using lentiviruses include low or absence of preexisting immunity, ability to transport one or more transgenes, delivery of genetic material to replicating and nonreplicating cells, as well as prolonged transgene expression (upward of 6 months). In order to make a lentivirus vector, a split component system is needed, where each part is in itself nonpathogenic and only the sum of it is parts can actively infect cells. Target cells are usually transfected with the viral vector, which is flanked with long terminal repeats (LTRs). It is this feature that allows the carried transgene to integrate into the genome of the target cell. The vector could also contain the Rev-responsive element (RRE) for most efficient vector production and, of course, the gene of interest. In parallel, a plasmid containing gag and pol structural genes are needed to supply reverse transcriptase and integration functions for the therapeutic vector particles. Finally, the last part is composed of plasmids encoding envelope proteins for the therapeutic viral particles and perhaps Rev. protein. Typically, envelope gene used is that of the glycoprotein G from vesicular stomatitis virus (VSV-G). The addition of this foreign viral envelope is called pseudotyping, and it alters the viral tropism to specifically target certain cell types.

Retroviruses and lentiviruses have been used to transfer genetic material since the 1980s. In the early 1990s, γ-retrovirus gene transfer was shown to be possible in hematopoietic stem cells [89]. This era also saw the first clinical trial that aimed at treating severe combined immunodeficiency (SCID) [90]. A major accomplishment in this field happened in the early 2000s, when 11 children were successfully treated for X-SCID by introducing the common interleukin receptor γ-chain in bone marrow using a retrovirus vector based on mouse leukemia virus (MLV) [91].

2.3.1. *Using lentiviruses for joint repair*

One of the first reports of using a lentivirus for the treatment of joints occurred in 2008. Ricchetti et al. overexpressed IL-10 in the patellar tendons of mice. IL-10 is known for its potent anti-inflammatory properties that limit host response to pathogens, but also can inhibit scar formation in fetal wound healing. In this study, a murine model of patellar tendon injury was used to investigate the effect of IL-10 overexpression on the properties of adult healing tendon. Findings showed successful transfer of IL-10 into patellar tendons with more than six times greater expression in comparison with endogenous IL-10 levels. IL-10 expression peaked at 10 days after injury. Furthermore, treated tendons showed improved maximum stress and percent relaxation was increased in the treated group. However, there were significant limitations regarding the study. The empty vector control also showed improved tendon properties compared to the sham control group, which could indicate that injection of the vector itself, rather than IL-10, as a beneficial effect. The authors hypothesize that injection of the viral vector may actually lead to more robust immune responses that subsequently drive better scar formation and wound healing.

2.3.2. *Lentiviruses toward cartilage regeneration*

Many attempts have been made to use retroviruses and lentiviruses for a long-term transgene expression in chondrocytes. Toward this, many different animal cells have been used, including human, rat, rabbit, goat, and cattle [92–95]. One group showed that transduction of chondrocytes with GFP was associated with an approximate 60% success rate [92]. After 6 weeks, only 21% of the cells remained GFP positive, whereas other studies showed greater efficiency rates with up to 85% of osteoarthritic chondrocytes being transduced [94]. Human articular chondrocytes have been shown to be highly susceptible to lentiviral infection, with 74% being GFP positive and expression was maintained *in vitro* for up to 22 weeks [93].

Like for the other viral vectors described in this chapter, studies have focused on inserting factors, which could help cartilage or bone repair, either by incorporating molecules stimulating the ECM, chondrogenesis, or immunomodulatory molecules. One such study examined the possibility of expressing a member of the nuclear factor of activated T-cells (NFAT) as a means to treat osteoarthritis [96]. NFAT was initially identified as a regulator of gene transcription in response to T-cell receptor-mediated signals in lymphocytes. However, it is also involved in regulating bone formation and osteoclastic bone resorption [97, 98]. Interestingly, NFAT knockout mice have normal skeletal development, but with age, display loss of type II collagen, and aggrecan. They also show overexpression of specific matrix-degrading proteinases including MMPs and ADAMTS in addition to proinflammatory cytokines. The authors then used a lentiviral vector to express NFAT1 in cultured primary *Nfat1*$^{-/-}$ articular chondrocytes. This rescue of NFAT partially or completely rescued the abnormal catabolic and anabolic activities of *Nfat1*$^{-/-}$ articular chondrocytes.

Another study looked at using the lentivirus vector to knock down aggrecanase activity [99]. RNAi was used to specifically target both aggrecanase-1 and -2 in primary rat chondrocytes. This approach was relatively successful *in vitro* with increased amounts of glycosaminoglycans and total collagen being produced as well as an increase in chondrocyte proliferation. This data provided the proof-of-principle that it is feasible to use this vector system to modulate chondrocyte phenotype and may be useful for future studies.

Several reports examined the ability of lentivirus vectors to be used to target MSCs in order to ameliorate the ECM surrounding the joints. One interesting example is the use of these vectors to help create a bioactive scaffold where sustained transgene expression and ECM formation are accomplished by human MSCs (hMSCs) [100]. The lentivirus vectors were used to express transforming growth factor β3 (TGF-β3) under the control of a constitutive EF-1α promoter. TGF-β3 was chosen as it was previously shown to be the most potent driver for chondrogenesis in hMSCs. After transduction, hMSCs developed a spherical shape comparable to chondrocyte-like morphology. Also, there was a substantial increase in col. II and glycosaminoglycan. Bioactive scaffolds with immobilized TGF-β3 expressed in lentivirus vectors showed a production of 17 ng/mL TGF-β3 and 12.87 μg sGAG/μg DNA at 1–3 weeks after seeding scaffolds. The results of this study indicate that the scaffold-mediated transduction technique could eventually be used *in vivo* to direct cell lineage commitment and ECM development in a controlled and persistent manner. The field of bioengineering is rapidly growing and the possibility of creating alternative methods for tissue replacement is not so far away.

One of the most recent publications examining the use of lentiviruses for cartilage repair used ovine perivascular stem cells (oPSCs). These cells are said to be natural ancestors of mesenchymal stem cells. The goal of this study was to develop an autologous large animal model for PSC transplantation and determine if implanted cells are retained in articular cartilage defects. oPSCs could be sourced from various locations including bone marrow, subcutaneous fat, and the infrapatellar fat pad. The lentivirus was used to transduce the cells with eGFP to allow tracking when implanted into the animals. The transduced cells were implanted into articular cartilage defects on the medial femoral condyle using hydrogel and collagen membranes. Results showed that GFP-emitting cells could be found at the base of the articular cartilage defect up to 4 weeks after transplantation. However, no repair tissue was seen by immunohistochemistry. Overall, more work needs to be done for this model to be a robust example of cartilage repair, but it could be an alternative replacement to the current canine model.

Despite some promising results, the use of lentiviruses will probably always raise concerns about safety due to the ability to integrate into the host genome. Clinical trials will be challenging due to the unknown risks associated with their administration. Thorough justification for their use will be warranted especially with so many other types of viral vectors currently available, although it is possible to see successful joint repair using such a system.

Overall, this chapter examined some of the most recent literature surrounding the use of viral vectors for bone and cartilage repair. This is a vast field with many exciting studies and promising developments. There has been a huge amount of progress since the early development of viral gene therapy, and it is only a matter of time before joint disorders and injuries will be treated using these approaches.

Author details

Penny A. Rudd and Lara J. Herrero*

*Address all correspondence to: l.herrero@griffith.edu.au

Institute for Glycomics, Griffith University, Southport, Qld, Australia

References

[1] Kirik D, Cederfjall E, Halliday G, Petersen A. Gene therapy for Parkinson's disease. Disease modification by GDNF family of ligands. Neurobiology of disease. 2017;**97**(Pt B):179-188

[2] Joshi CR, Labhasetwar V, Ghorpade A. Destination brain: The past, present, and future of therapeutic gene delivery. Journal of Neuroimmune Pharmacology: The Official Journal of the Society on Neuroimmune Pharmacology. 2017;**12**(1):51-83

[3] Kumar SR, Markusic DM, Biswas M, High KA, Herzog RW. Clinical development of gene therapy: Results and lessons from recent successes. Molecular Therapy Methods & Clinical Development. 2016;**3**:16034

[4] Simonato M, Bennett J, Boulis NM, Castro MG, Fink DJ, Goins WF, et al. Progress in gene therapy for neurological disorders. Nature Reviews Neurology. 2013;**9**(5):277-291

[5] Yla-Herttuala S, Baker AH. Cardiovascular gene therapy: Past, present, and future. Molecular Therapy: The Journal of the American Society of Gene Therapy. 2017;**25**(5):1095-1106

[6] Evans CH, Huard J. Gene therapy approaches to regenerating the musculoskeletal system. Nature Reviews Rheumatology. 2015;**11**(4):234-242

[7] Loeser RF. Molecular mechanisms of cartilage destruction: Mechanics, inflammatory mediators, and aging collide. Arthritis and Rheumatism. 2006;**54**(5):1357-1360

[8] Buckwalter JA, Brown TD. Joint injury, repair, and remodeling: Roles in post-traumatic osteoarthritis. Clinical orthopaedics and related research. 2004;**423**:7-16

[9] Evans CH, Kraus VB, Setton LA. Progress in intra-articular therapy. Nature Reviews Rheumatology. 2014;**10**(1):11-22

[10] McAlindon TE, Bannuru RR, Sullivan MC, Arden NK, Berenbaum F, Bierma-Zeinstra SM, et al. OARSI guidelines for the non-surgical management of knee osteoarthritis. Osteoarthritis and cartilage. 2014;**22**(3):363-388

[11] van der Hoek L, Pyrc K, Jebbink MF, Vermeulen-Oost W, Berkhout RJ, Wolthers KC, et al. Identification of a new human coronavirus. Nature Medicine. 2004;**10**(4):368-373.

[12] Woo PC, Lau SK, Chu CM, Chan KH, Tsoi HW, Huang Y, et al. Characterization and complete genome sequence of a novel coronavirus, coronavirus HKU1, from patients with pneumonia. Journal of Virology. 2005;**79**(2):884-895

[13] Chan JF, Lau SK, To KK, Cheng VC, Woo PC, Yuen KY. Middle East respiratory syndrome coronavirus: Another zoonotic betacoronavirus causing SARS-like disease. Clinical Microbiology Reviews. 2015;**28**(2):465-522

[14] Perlman S, Netland J. Coronaviruses post-SARS: Update on replication and pathogenesis. Nature Reviews Microbiology. 2009;**7**(6):439-450

[15] Enjuanes L, Sola I, Almazan F, Ortego J, Izeta A, Gonzalez JM, et al. Coronavirus derived expression systems. Journal of Biotechnology. 2001;**88**(3):183-204

[16] Alonso S, Sola I, Teifke JP, Reimann I, Izeta A, Balasch M, et al. Vitro and in vivo expression of foreign genes by transmissible gastroenteritis coronavirus-derived minigenomes. The Journal of General Virology. 2002;**83**(Pt 3):567-579

[17] de Haan CA, van Genne L, Stoop JN, Volders H, Rottier PJ. Coronaviruses as vectors: Position dependence of foreign gene expression. Journal of Virology. 2003;**77**(21):11312-11323

[18] Cervantes-Barragan L, Zust R, Maier R, Sierro S, Janda J, Levy F, et al. Dendritic cell-specific antigen delivery by coronavirus vaccine vectors induces long-lasting protective antiviral and antitumor immunity. MBio. 2010;**1**:e00171-10. doi: 10.1128/mBio.00171-10

[19] Shen ZT, Sigalov AB. SARS coronavirus fusion peptide-derived sequence suppresses collagen-induced arthritis in DBA/1J mice. Scientific Reports. 2016;**6**:28672

[20] Bowles DE, Rabinowitz JE, Samulski RJ. Parvoviruses. In: Kerr JR, Cotmore SF, Bloom ME, Linden RM, Parrish CR, editors. Parvoviruses. New York: Edward Arnold Ltd.; 2006. p. 15-24

[21] Bueler H. Adeno-associated viral vectors for gene transfer and gene therapy. Biological Chemistry. 1999;**380**(6):613-622

[22] Blacklow NR, Hoggan MD, Rowe WP. Isolation of adenovirus-associated viruses from man. Proceedings of the National Academy of Sciences of the United States of America. 1967;**58**(4):1410-1415

[23] Flotte TR, Barraza-Ortiz X, Solow R, Afione SA, Carter BJ, Guggino WB. An improved system for packaging recombinant adeno-associated virus vectors capable of in vivo transduction. Gene Therapy. 1995;**2**(1):29-37

[24] Tenenbaum L, Humbert-Claude M. Glial cell line-derived Neurotrophic factor gene delivery in Parkinson's disease: A delicate balance between neuroprotection, trophic effects, and unwanted compensatory mechanisms. Frontiers in Neuroanatomy. 2017;**11**:29

[25] Loring HS, Flotte TR. Current status of gene therapy for alpha-1 antitrypsin deficiency. Expert opinion on biological therapy. 2015;**15**(3):329-336

[26] Ward P, Walsh CE. Current and future prospects for hemophilia gene therapy. Expert Review of Hematology. 2016;**9**(7):649-659

[27] Smith BK, Collins SW, Conlon TJ, Mah CS, Lawson LA, Martin AD, et al. Phase I/II trial of adeno-associated virus-mediated alpha-glucosidase gene therapy to the diaphragm for chronic respiratory failure in Pompe disease: Initial safety and ventilatory outcomes. Human Gene Therapy. 2013;**24**(6):630-640

[28] Mingozzi F, High KA. Therapeutic in vivo gene transfer for genetic disease using AAV: Progress and challenges. Nature Reviews Genetics. 2011;**12**(5):341-355

[29] Bryant LM, Christopher DM, Giles AR, Hinderer C, Rodriguez JL, Smith JB, et al. Lessons learned from the clinical development and market authorization of Glybera. Human Gene Therapy Clinical development. 2013;**24**(2):55-64

[30] Russell DW, Miller AD, Alexander IE. Adeno-associated virus vectors preferentially transduce cells in S phase. Proceedings of the National Academy of Sciences of the United States of America. 1994;**91**(19):8915-8919

[31] Santangelo KS, Bertone AL. Effective reduction of the interleukin-1beta transcript in osteoarthritis-prone guinea pig chondrocytes via short hairpin RNA mediated RNA interference influences gene expression of mediators implicated in disease pathogenesis. Osteoarthritis and Cartilage. 2011;**19**(12):1449-1457

[32] Watson RS, Broome TA, Levings PP, Rice BL, Kay JD, Smith AD, et al. scAAV-mediated gene transfer of interleukin-1-receptor antagonist to synovium and articular cartilage in large mammalian joints. Gene therapy. 2013;**20**(6):670-677

[33] Goodrich LR, Phillips JN, McIlwraith CW, Foti SB, Grieger JC, Gray SJ, et al. Optimization of scAAVIL-1ra in vitro and in vivo to deliver high levels of therapeutic protein for treatment of osteoarthritis. Molecular Therapy Nucleic Acids. 2013;**2** e70

[34] Nayak S, Herzog RW. Progress and prospects: Immune responses to viral vectors. Gene Therapy. 2010;**17**(3):295-304

[35] Mueller C, Flotte TR. Clinical gene therapy using recombinant adeno-associated virus vectors. Gene Therapy. 2008;**15**(11):858-863

[36] Vosse D, de Vlam K. Osteoporosis in rheumatoid arthritis and ankylosing spondylitis. Clinical and Experimental Rheumatology. 2009;**27**(4 Suppl 55):S62-S67

[37] Kocijan R, Englbrecht M, Haschka J, Simon D, Kleyer A, Finzel S, et al. Quantitative and qualitative changes of bone in psoriasis and psoriatic arthritis patients. Journal of Bone And Mineral Research: The Official Journal of the American Society for Bone and Mineral Research. 2015;**30**(10):1775-1783

[38] Manolagas SC, Jilka RL. Bone marrow, cytokines, and bone remodeling. Emerging insights into the pathophysiology of osteoporosis. The New England Journal of Medicine. 1995;**332**(5):305-311

[39] Katagiri T, Yamaguchi A, Komaki M, Abe E, Takahashi N, Ikeda T, et al. Bone morphogenetic protein-2 converts the differentiation pathway of C2C12 myoblasts into the osteoblast lineage. The Journal of Cell Biology. 1994;**127**(6 Pt 1):1755-1766

[40] Luk KD, Chen Y, Cheung KM, Kung HF, WW L, Leong JC. Adeno-associated virus-mediated bone morphogenetic protein-4 gene therapy for in vivo bone formation. Biochemical and Biophysical Research Communications. 2003;**308**(3):636-645

[41] Chen Y, Luk KD, Cheung KM, Xu R, Lin MC, WW L, et al. Gene therapy for new bone formation using adeno-associated viral bone morphogenetic protein-2 vectors. Gene Therapy. 2003;**10**(16):1345-1353

[42] Ben Arav A, Pelled G, Zilberman Y, Kimelman-Bleich N, Gazit Z, Schwarz EM, et al. Adeno-associated virus-coated allografts: A novel approach for cranioplasty. Journal of Tissue Engineering And Regenerative Medicine. 2012;**6**(10):e43-e50

[43] Koefoed M, Ito H, Gromov K, Reynolds DG, Awad HA, Rubery PT, et al. Biological effects of rAAV-caAlk2 coating on structural allograft healing. Molecular Therapy: The Journal Of the American Society of Gene Therapy. 2005;**12**(2):212-218

[44] Ito H, Koefoed M, Tiyapatanaputi P, Gromov K, Goater JJ, Carmouche J, et al. Remodeling of cortical bone allografts mediated by adherent rAAV-RANKL and VEGF gene therapy. Nature Medicine. 2005;**11**(3):291-297

[45] Colnot C, Thompson Z, Miclau T, Werb Z, Helms JA. Altered fracture repair in the absence of MMP9. Development. 2003;**130**(17):4123-4133

[46] Kon T, Cho TJ, Aizawa T, Yamazaki M, Nooh N, Graves D, et al. Expression of osteoprotegerin, receptor activator of NF-kappaB ligand (osteoprotegerin ligand) and related proinflammatory cytokines during fracture healing. Journal of Bone and Mineral Research: The Official Journal of the American Society for Bone and Mineral Research. 2001;**16**(6):1004-1014

[47] Ferrara N, Gerber HP, LeCouter J. The biology of VEGF and its receptors. Nature Medicine. 2003;**9**(6):669-676

[48] Griffin DJ, Ortved KF, Nixon AJ, Bonassar LJ. Mechanical properties and structure-function relationships in articular cartilage repaired using IGF-I gene-enhanced chondrocytes. Journal of Orthopaedic Research: Official Publication of the Orthopaedic Research Society. 2016;**34**(1):149-153

[49] Cucchiarini M, Schetting S, Terwilliger EF, Kohn D, Madry H. rAAV-mediated overexpression of FGF-2 promotes cell proliferation, survival, and alpha-SMA expression in human meniscal lesions. Gene Therapy. 2009;**16**(11):1363-1372

[50] Zhang W, Wang F, Wang B, Zhang J, Yu JY. Intraarticular gene delivery of CTLA4-FasL suppresses experimental arthritis. International Immunology. 2012;**24**(6):379-388

[51] Kyostio-Moore S, Bangari DS, Ewing P, Nambiar B, Berthelette P, Sookdeo C, et al. Local gene delivery of heme oxygenase-1 by adeno-associated virus into osteoarthritic mouse joints exhibiting synovial oxidative stress. Osteoarthritis and Cartilage. 2013;**21**(2):358-367

[52] Tas SW, Adriaansen J, Hajji N, Bakker AC, Firestein GS, Vervoordeldonk MJ, et al. Amelioration of arthritis by intraarticular dominant negative Ikk beta gene therapy using adeno-associated virus type 5. Human Gene Therapy. 2006;**17**(8):821-832

[53] Cucchiarini M, Terwilliger EF, Kohn D, Madry H. Remodelling of human osteoarthritic cartilage by FGF-2, alone or combined with Sox9 via rAAV gene transfer. Journal of Cellular and Molecular Medicine. 2009;**13**(8B):2476-2488

[54] Bao L, Zhu T, Zhao D, Han X, Guan J, Shi Z, et al. Adeno-associated virus-mediated osteoprotegerin gene transfer protects against joint destruction in a collagen-induced arthritis rat model. Joint, Bone, Spine: Revue du Rhumatisme. 2012;**79**(5):482-487

[55] Venkatesan JK, Frisch J, Rey-Rico A, Schmitt G, Madry H, Cucchiarini M. Impact of mechanical stimulation on the chondrogenic processes in human bone marrow aspirates modified to overexpress sox9 via rAAV vectors. Journal of Experimental Orthopaedics. 2017;**4**(1):22

[56] Wang C, Ruan DK, Zhang C, Wang DL, Xin H, Zhang Y. Effects of adeno-associated virus-2-mediated human BMP-7 gene transfection on the phenotype of nucleus pulposus cells. Journal of Orthopaedic Research: Official Publication of the Orthopaedic Research Society. 2011;**29**(6):838-845

[57] Boutet MA, Najm A, Bart G, Brion R, Touchais S, Trichet V, et al. IL-38 overexpression induces anti-inflammatory effects in mice arthritis models and in human macrophages in vitro. Annals of the Rheumatic Diseases. 2017;**76**(7):1304-1312

[58] Kyostio-Moore S, Piraino S, Berthelette P, Moran N, Serriello J, Bendele A, et al. Overexpression of cystatin C in synovium does not reduce synovitis or cartilage degradation in established osteoarthritis. Arthritis Research & Therapy. 2015;**17**:5

[59] Arai Y, Kubo T, Fushiki S, Mazda O, Nakai H, Iwaki Y, et al. Gene delivery to human chondrocytes by an adeno associated virus vector. The Journal of Rheumatology. 2000;**27**(4):979-982

[60] Rohr UP, Kronenwett R, Grimm D, Kleinschmidt J, Haas R. Primary human cells differ in their susceptibility to rAAV-2-mediated gene transfer and duration of reporter gene expression. Journal of Virological Methods. 2002;**105**(2):265-275

[61] Cucchiarini M, Madry H, Ma C, Thurn T, Zurakowski D, Menger MD, et al. Improved tissue repair in articular cartilage defects in vivo by rAAV-mediated overexpression of human fibroblast growth factor 2. Molecular Therapy: The Journal of the American Society of Gene Therapy. 2005;**12**(2):229-238

[62] Chamberlain JR, Schwarze U, Wang PR, Hirata RK, Hankenson KD, Pace JM, et al. Gene targeting in stem cells from individuals with osteogenesis imperfecta. Science. 2004; **303**(5661):1198-1201

[63] Ulrich-Vinther M, Maloney MD, Goater JJ, Soballe K, Goldring MB, O'Keefe RJ, et al. Light-activated gene transduction enhances adeno-associated virus vector-mediated gene expression in human articular chondrocytes. Arthritis and Rheumatism. 2002;**46**(8):2095-2104

[64] Ulrich-Vinther M, Duch MR, Soballe K, O'Keefe RJ, Schwarz EM, Pedersen FS. Vivo gene delivery to articular chondrocytes mediated by an adeno-associated virus vector. Journal of Orthopaedic Research: Official Publication of the Orthopaedic Research Society. 2004;**22**(4):726-734

[65] Baragi VM, Renkiewicz RR, Jordan H, Bonadio J, Hartman JW, Roessler BJ. Transplantation of transduced chondrocytes protects articular cartilage from interleukin 1-induced extracellular matrix degradation. The Journal of Clinical Investigation. 1995;**96**(5):2454-2460

[66] Nixon AJ, Brower-Toland BD, Bent SJ, Saxer RA, Wilke MJ, Robbins PD, et al. Insulinlike growth factor-I gene therapy applications for cartilage repair. Clinical Orthopaedics and Related Research. 2000;**379**(Suppl):S201-S213

[67] Danthinne X, Imperiale MJ. Production of first generation adenovirus vectors: A review. Gene Therapy. 2000;**7**(20):1707-1714

[68] Alba R, Bosch A, Chillon M. Gutless adenovirus: Last-generation adenovirus for gene therapy. Gene Therapy. 2005;**12**(Suppl 1):S18-S27

[69] Roessler BJ, Allen ED, Wilson JM, Hartman JW, Davidson BL. Adenoviral-mediated gene transfer to rabbit synovium in vivo. The Journal of Clinical Investigation. 1993;**92**(2):1085-1092

[70] Roessler BJ, Hartman JW, Vallance DK, Latta JM, Janich SL, Davidson BL. Inhibition of interleukin-1-induced effects in synoviocytes transduced with the human IL-1 receptor antagonist cDNA using an adenoviral vector. Human Gene Therapy. 1995;**6**(3):307-316

[71] Arend WP, Malyak M, Guthridge CJ, Gabay C. Interleukin-1 receptor antagonist: Role in biology. Annual Review of Immunology. 1998;**16**:27-55

[72] Kang Q, Sun MH, Cheng H, Peng Y, Montag AG, Deyrup AT, et al. Characterization of the distinct orthotopic bone-forming activity of 14 BMPs using recombinant adenovirus-mediated gene delivery. Gene Therapy. 2004;**11**(17):1312-1320

[73] Urist MR. Bone: Formation by autoinduction. Science. 1965;**150**(3698):893-899

[74] Poynton AR, Lane JM. Safety profile for the clinical use of bone morphogenetic proteins in the spine. Spine. 2002;**27**(16 Suppl 1):S40-S48

[75] Sonobe J, Okubo Y, Kaihara S, Miyatake S, Bessho K. Osteoinduction by bone morphogenetic protein 2-expressing adenoviral vector: Application of biomaterial to mask the host immune response. Human Gene Therapy. 2004;**15**(7):659-668

[76] Franceschi RT, Wang D, Krebsbach PH, Rutherford RB. Gene therapy for bone formation: In vitro and in vivo osteogenic activity of an adenovirus expressing BMP7. Journal of Cellular Biochemistry. 2000;**78**(3):476-486

[77] Sawchuk SJ, Boivin GP, Duwel LE, Ball W, Bove K, Trapnell B, et al. Anti-T cell receptor monoclonal antibody prolongs transgene expression following adenovirus-mediated in vivo gene transfer to mouse synovium. Human Gene Therapy. 1996;**7**(4):499-506

[78] Cao H, Sun ZB, Zhang L, Qian W, Li CY, Guo XP, et al. Adenovirus-mediated bone morphogenetic protein-2 promotes osteogenic differentiation in human mesenchymal stem cells in vitro. Experimental and Therapeutic Medicine. 2017;**14**(1):377-382

[79] Zhang X, Kuroda S, Carpenter D, Nishimura I, Soo C, Moats R, et al. Craniosynostosis in transgenic mice overexpressing Nell-1. The Journal of Clinical Investigation. 2002;**110**(6):861-870

[80] Aghaloo T, Cowan CM, Chou YF, Zhang X, Lee H, Miao S, et al. Nell-1-induced bone regeneration in calvarial defects. The American Journal of Pathology. 2006;**169**(3):903-915

[81] Chen Y, Caporali E, Stewart M. Bone morphogenetic protein 2 stimulates chondrogenesis of equine synovial membrane-derived progenitor cells. Veterinary and Comparative Orthopaedics and Traumatology: VCOT. 2016;**29**(5):378-385

[82] Goodrich LR, Hidaka C, Robbins PD, Evans CH, Nixon AJ. Genetic modification of chondrocytes with insulin-like growth factor-1 enhances cartilage healing in an equine model. The Journal of Bone and Joint Surgery British volume. 2007;**89**(5):672-685

[83] Chen B, Qin J, Wang H, Magdalou J, Chen L. Effects of adenovirus-mediated bFGF, IL-1Ra and IGF-1 gene transfer on human osteoarthritic chondrocytes and osteoarthritis in rabbits. Experimental & Molecular Medicine. 2010;**42**(10):684-695

[84] Haupt JL, Frisbie DD, McIlwraith CW, Robbins PD, Ghivizzani S, Evans CH, et al. Dual transduction of insulin-like growth factor-I and interleukin-1 receptor antagonist protein controls cartilage degradation in an osteoarthritic culture model. Journal of Orthopaedic Research: Official Publication of the Orthopaedic Research Society. 2005;**23**(1):118-126

[85] Luo XW, Liu K, Chen Z, Zhao M, Han XW, Bai YG, et al. Adenovirus-mediated GDF-5 promotes the extracellular matrix expression in degenerative nucleus pulposus cells. Journal of Zhejiang University Science B. 2016;**17**(1):30-42

[86] Wright E, Hargrave MR, Christiansen J, Cooper L, Kun J, Evans T, et al. The Sry-related gene Sox9 is expressed during chondrogenesis in mouse embryos. Nature Genetics. 1995;**9**(1): 15-20

[87] Cao L, Yang F, Liu G, Yu D, Li H, Fan Q, et al. The promotion of cartilage defect repair using adenovirus mediated Sox9 gene transfer of rabbit bone marrow mesenchymal stem cells. Biomaterials. 2011;**32**(16):3910-3920

[88] Sibbald B. Death but one unintended consequence of gene-therapy trial. CMAJ: Canadian Medical Association journal = journal de l'Association medicale canadienne. 2001;**164**(11):1612

[89] Brenner MK, Rill DR, Holladay MS, Heslop HE, Moen RC, Buschle M, et al. Gene marking to determine whether autologous marrow infusion restores long-term haemopoiesis in cancer patients. Lancet. 1993;**342**(8880):1134-1137

[90] Anderson WF, Blaese RM, Culver K. The ADA human gene therapy clinical protocol: Points to consider response with clinical protocol 06-07-1990. Human Gene Therapy. 1990;**1**(3):331-362

[91] Cavazzana-Calvo M, Hacein-Bey S, de Saint Basile G, Gross F, Yvon E, Nusbaum P, et al. Gene therapy of human severe combined immunodeficiency (SCID)-X1 disease. Science. 2000;**288**(5466):669-672

[92] FZ L, Kitazawa Y, Hara Y, Jiang JY, Li XK. Long-term gene expression using the lentiviral vector in rat chondrocytes. Clinical Orthopaedics and Related Research. 2005;**439**:243-252

[93] Miot S, Gianni-Barrera R, Pelttari K, Acharya C, Mainil-Varlet P, Juelke H, et al. In vitro and in vivo validation of human and goat chondrocyte labeling by green fluorescent protein lentivirus transduction. Tissue Engineering Part C, Methods. 2010;**16**(1):11-21

[94] Li Y, Tew SR, Russell AM, Gonzalez KR, Hardingham TE, Hawkins RE. Transduction of passaged human articular chondrocytes with adenoviral, retroviral, and lentiviral vectors and the effects of enhanced expression of SOX9. Tissue Engineering. 2004;**10**(3-4):575-584

[95] Hirschmann F, Verhoeyen E, Wirth D, Bauwens S, Hauser H, Rudert M. Vital marking of articular chondrocytes by retroviral infection using green fluorescence protein. Osteoarthritis and Cartilage. 2002;**10**(2):109-118

[96] Wang J, Gardner BM, Lu Q, Rodova M, Woodbury BG, Yost JG, et al. Transcription factor Nfat1 deficiency causes osteoarthritis through dysfunction of adult articular chondrocytes. The Journal of Pathology. 2009;**219**(2):163-172

[97] Koga T, Matsui Y, Asagiri M, Kodama T, de Crombrugghe B, Nakashima K, et al. NFAT and Osterix cooperatively regulate bone formation. Nature Medicine.. 2005;**11**(8):880-885

[98] Takayanagi H, Kim S, Koga T, Nishina H, Isshiki M, Yoshida H, et al. Induction and activation of the transcription factor NFATc1 (NFAT2) integrate RANKL signaling in terminal differentiation of osteoclasts. Developmental Cell. 2002;**3**(6):889-901

[99] Wang ZH, Yang ZQ, He XJ, Kamal BE, Xing Z. Lentivirus-mediated knockdown of aggrecanase-1 and -2 promotes chondrocyte-engineered cartilage formation in vitro. Biotechnology and Bioengineering. 2010;**107**(4):730-736

[100] Brunger JM, Huynh NP, Guenther CM, Perez-Pinera P, Moutos FT, Sanchez-Adams J, et al. Scaffold-mediated lentiviral transduction for functional tissue engineering of cartilage. Proceedings of the National Academy of Sciences of the United States of America. 2014;**111**(9):E798-E806

Chapter 2

Chondrocyte Turnover in Lung Cartilage

Yareth Gopar-Cuevas, Alberto Niderhauser-García,
Adriana Ancer-Arellano,
Ivett C. Miranda-Maldonado,
María-de-Lourdes Chávez-Briones,
Laura E. Rodríguez-Flores,
Marta Ortega-Martínez and
Gilberto Jaramillo-Rangel

Additional information is available at the end of the chapter

http://dx.doi.org/10.5772/intechopen.70860

Abstract

Cartilage is a highly differentiated connective tissue that forms mechanical support to soft tissues and is important for bone development from fetal period to puberty. It is conformed by chondrocytes and extracellular matrix. It is generally believed that adult cartilage has no capacity to renewal. A delicate balance between cell proliferation and cell death ensures the maintenance of normal tissue morphology and function. Stem cells play essential roles in this process. Mesenchymal stem cells (MSCs) can give rise to multiple lineages including bone, adipose and cartilage. Nestin protein was initially identified as a marker for neural stem cells, but its expression has also been detected in many types of cells, including MSCs. *In vivo*, chondrocyte turnover has been almost exclusively studied in articular cartilage. In this chapter we will review the findings about the chondrocyte turnover in lung cartilage. We have presented evidence that there exist nestin-positive MSCs in healthy adulthood that participates in the turnover of lung cartilage and in lung airway epithelium renewal. These findings may improve our knowledge about the biology of the cartilage and of the stem cells, and could provide new cell candidates for cartilage tissue engineering and for therapy for devastating pulmonary diseases.

Keywords: lung, cartilage, chondrocyte, turnover, apoptosis, proliferation, stem cells

1. Introduction

Cartilage is a connective tissue consisting of cells and extracellular matrix (ECM). These cells are called chondrocytes and reside within spaces called lacunae. The ECM is a three-dimensional macromolecular network composed of fibers and ground substance. In the mammals, much of the skeleton is first laid down in cartilage, and in the adult body it covers the articular surfaces of bones and forms the sole skeletal support of several structures [1].

Normal chondrocytes maintain a functional ECM that replaces itself very slowly and provides a shock absorber [2]. However, in mature cartilage, metabolic activity is low, and has been thought that adult chondrocytes resist proliferation throughout life. As a result, the mechanical properties of cartilage deteriorate with age [3–5].

Cell death and cell proliferation must be balanced in adult organisms in order to maintain homeostasis. Programmed cell death or apoptosis is important in mature organisms for deleting unwanted cells (e.g. aged cells). Most tissues contain stem cells that are able of proliferate and differentiate to replace cells that have been lost. A defective cell turnover process may have serious consequences to the tissues and the entire organism [6].

The role of chondrocyte turnover in cartilage aging and disease has been poorly analyzed and most of the related studies have been carried out in articular cartilage. In this chapter we will review the findings about the chondrocyte turnover in lung cartilage.

2. Chondrocyte, cartilage, and pulmonary cartilage

2.1. The chondrocyte

There are two forms of cells in cartilage: chondroblasts and chondrocytes. Chondroblasts are actively dividing immature cells which form ECM. They are oval or spindle-shaped cells with a spherical nucleus. The cytoplasm is basophilic, rich in ribosomes, rough endoplasmic reticulum, and Golgi saccules [7].

When chondroblasts are completely surrounded by ECM, they are called chondrocytes. They reside in spaces within the cartilage matrix known as lacunae. However, the cells fill the lacunae *in vivo*, as verified by electron microscopic studies. Chondrocytes vary from elongate to spherical in shape in relation to their position within the cartilage. They have a spherical nucleus with one or more nucleoli. Chondrocyte cytoplasm contains, in addition to glycogen and lipid, the usual characteristics of a secretory cell: abundant rough endoplasmic reticulum and prominent Golgi complex [8, 9].

The main function of the chondrocyte is to produce, maintain, and remodel the ECM of the cartilage. Chondrocytes receive mechanical, electrical, and physicochemical signals transmitted by the ECM and respond by regulating their metabolic activity [3, 9].

2.2. Cartilage

Cartilage is flexible and strong, and is resilient to compression. It forms mechanical support to soft tissues and is important for bone development from fetal period to puberty [1, 10].

Cartilage consists of cells (chondroblasts and chondrocytes) and ECM. The ECM is primarily composed of tissue fluid and macromolecules, including collagens, proteoglycans, and glycoproteins. Cartilage is subdivided into three varieties depending on their molecular composition: hyaline, elastic, and fibrous [11]. Of these, hyaline cartilage is the most widely distributed type.

With the exception of the free surfaces of articular cartilages, hyaline cartilage is surrounded by a membrane of fibrous connective tissue, the perichondrium. Cartilage is usually devoid of blood vessels, so its cells must obtain their oxygen and nutrients by long-range diffusion from the perichondrium [12].

2.3. Pulmonary cartilage

The upper respiratory tract includes the nose and nasal passages, paranasal sinuses, the pharynx, and the portion of the larynx above the vocal cords. The lower respiratory tract includes the trachea and within the lungs, the bronchi, bronchioles, and alveoli. This system performs or participates in several functions: air conduction, gas exchange, olfaction, and phonation [13–15].

Although the air passages take on their mature appearance well before a fetus is viable, they undergo significant maturational changes in late gestation. Thereafter, the lungs undergo a phase of growth and maturation during the first two decades of live and achieve maximal lung function approximately at the age of 20 years old for women and 25 years old for men. Lung function remains steady from age 20 to 35 years and starts declining thereafter. It has been suggested that airway cartilage plays an important role in determining airway compressibility and distensibility. Age-related differences in airway mechanical function may reflect an increase in stiffness of both airway muscle and cartilage that occurs with increasing age [16–18].

Cartilage (hyaline type) has the function of maintaining airway patency and it also serves for the attachment of local muscle and connective tissue. It exists in the form of plates of cartilage which have characteristics shapes and arrangements at different airway levels [19].

In the trachea and right and left main bronchi, cartilage is present in the anterior and lateral walls as C-shaped plates. Approximately 15–20 cartilaginous rings support the trachea. The cartilage in the wall of intrapulmonary bronchi is in the form of irregular cartilage plates that form a complete but not continuous circumferential support. The smallest bronchi have only widely scattered cartilaginous plates in their walls. Terminal and respiratory bronchioles lack supporting cartilaginous plates [13, 19].

3. Cell turnover

Physiological cell turnover plays an important role in maintaining normal tissue function and morphology. During this process, older differentiated cells are typically eliminated by programmed cell death (apoptosis) and replaced by the division progeny of adult stem cells (ASC) [20, 21].

A delicate balance among all factors influencing cell turnover is needed to maintain the normal volume and function of tissues in healthy people. The key points of this homoeostatic process are apoptosis and cell proliferation. Cell turnover is precisely regulated by the interplay of various factors, which modulate tissue and cell-specific responses on apoptosis and proliferation, either directly, or by altering expression and function of key death and/or cell proliferative genes [6, 20, 22].

Age-specific changes in tissue regeneration and repair lead to cell loss and compromise of tissue homeostasis, structure, and function. These phenomena parallel changes in resident stem cell function [23, 24].

3.1. Apoptosis

Apoptosis is a process of controlled cellular death whereby the activation of specific death-signaling pathways leads to deletion of cells from tissue [25]. The term apoptosis was first used in a paper by Kerr, Wyllie, and Currie in 1972 to describe a morphologically distinct form of cell death [26], discriminating it from necrosis.

Apoptosis plays an essential role in survival of the organisms and is responsible for many biological processes such as normal cell turnover, embryonic and brain development, proper development and functioning of the immune system, and hormone-dependent atrophy [27, 28].

3.1.1. Apoptosis versus necrosis. Other forms of cell death

Cell death has been broadly classified in two categories: apoptosis and necrosis. Apoptosis is a synchronized and energy-requiring process than involves altered expression of key cell proliferation and death-inducing genes, and the activation of a group of cysteine proteases (caspases) in a complex cascade of events that link the initiating stimuli to the final demise of the cell, while necrosis does not involve gene expression and is a passive externally driven event resulting from acute cellular injury [20, 29]. However, increasing evidence has been accumulating that necrosis can occur in a regulated manner, and that necrosis has a prominent role in multiple physiological and pathological settings [30].

Apoptosis is morphologically characterized by cell shrinkage, detachment from the substrate, chromatin condensation, nuclear and DNA fragmentation, cytoplasmic membrane blebbing, package of the cell debris into apoptotic bodies, and engulfment by resident phagocytes. Necrosis involves increase in cell volume, swelling of organelles, rupture of the plasma membrane, and the subsequent release of the cytoplasmic contents into the surrounding tissue, leading to inflammatory reaction [31].

Recently, new forms of cell death have been progressively described, which can be more precisely distinguished based on molecular pathways. A functional classification of cell death forms have been proposed that includes extrinsic apoptosis, caspase-dependent or -independent intrinsic apoptosis, regulated necrosis, autophagic cell death, and mitotic catastrophe [30, 31].

3.1.2. Apoptosis mechanisms

Apoptosis can be initiated by exogenous stimuli such as ionizing radiation and chemotherapeutic drugs, as well as by endogenous stimuli such as the absence of oxygen, nutrients or growth/survival factors, the presence of DNA damage, or the action of cytokines [32].

There are two main apoptotic pathways: the extrinsic or death receptor pathway, which is triggered from outside of the cell by death ligands, and the intrinsic or mitochondrial pathway, which is triggered from inside the cell as a response to various stress signals. Both intrinsic as well as extrinsic pathways of apoptosis are associated and influence each other [33]. Another pathway of apoptosis as also been recognized that involves T- and NK-cell mediated cytotoxicity and perforin-granzyme-dependent killing of the cell [34].

The three pathways converge on the same execution pathway: the activation of cysteine proteases of the caspase family, which selectively digest the cell from within. The perforin/granzyme pathway also activates another cell death pathway via single stranded DNA damage [29, 34].

3.1.3. *Methods of apoptosis detection*

Since the pathways of apoptosis are very complicated, there are a lot of features of it than can be evaluated. A great number of methods have been developed to detect apoptosis, such as morphological techniques, proteomic and genomic approaches, spectroscopic methods, flow cytometry, caspase activity assays, microfluidic applications, and electrochemical methods [35]. Each assay has advantages and disadvantages. Understanding the strengths and limitations of the assays would allow investigators to select the best methods for their needs [28, 36]. A description of all assays for detecting apoptosis is beyond the scope of this chapter. We will briefly describe the assays to detect apoptosis most employed by our group.

3.1.3.1. *Light microscopy*

Detection of apoptotic cells in hematoxylin and eosin-stained tissue sections with light microscopy is possible because of characteristic morphological features of apoptosis. They include condensation of the chromatin in granular masses along the nuclear envelope, cell shrinkage, convolution of the cellular and nuclear outlines, and fragmentation of the nucleus. The apoptotic cell breaks into membrane bound bodies that are quickly removed by neighboring macrophages. The condensed or fragmented nucleus can be detected with DNA dyes such as propidium iodide, Hoechst dye, or DAPI (4′,6-diamidino-2-phenylindole). Light microscopy detects the later events of apoptosis and confirmation with other methods may be necessary [37, 38].

3.1.3.2. *Transmission electron microscopy (TEM)*

A more definitive method of morphologic identification of apoptotic cells is TEM, because apoptosis is confirmed by several of its ultrastructural characteristics. TEM detects chromatin condensation and convulsions in and around the nuclear envelope that precedes nuclear fragmentation, the condensation of cytoplasm with the disappearance of the microvilli, blebs on the cell surface, and the loss of cell junctions. If immunochemical staining is employed, then chemical information can be also obtained. However, there are limitations in TEM as an apoptosis detection method, including that apoptotic cells detected by TEM are in the last stage of apoptosis, and that much time and a high skill are required for preparation of ultrathin sections used in TEM [35, 39].

3.1.3.3. Terminal deoxynucleotidyl transferase-mediated dUTP nick end-labeling (TUNEL)

TUNEL method is based on the assumption that genomic DNA is fragmented in a dying cell, producing fragments of consistent length in apoptotic cell death, as opposed to necrotic cell death where DNA is believed to be randomly degraded [40, 41]. The method consists of the labeling of DNA nick ends by terminal deoxynucleotidyl transferase (TdT) which incorporates the labeled nucleotide (most often dUTP) in the places of DNA strain breaks. The dUTP can then be labeled with a variety of probes to allow detection by light microscopy, fluorescence microscopy, or flow cytometry [42].

TUNEL method is suitable for analysis of apoptosis in individual cells applicable to all kinds of material: cultured cells, tissues, and blood samples, even if a material contains only a few apoptotic cells. Another advantage of the TUNEL staining is that detects cells at a relatively early stage of apoptosis [39, 43]. However, this method also has drawbacks. Notably, it has been reported that the TUNEL assay also detect necrotic and autolytic cells in addition to apoptotic cells [44, 45].

3.2. Cell proliferation

Cell proliferation is the process whereby cells reproduce themselves by growing and then dividing into two equal copies [46]. This process is a fundamental requirement for normal development and homeostasis.

Cell division consists of two consecutive processes, mainly characterized by DNA replication and segregation of replicated chromosomes into two separate cells. The process of replicating DNA and dividing a cell can be described as a series of coordinated events that compose a cell cycle [47, 48].

3.2.1. The cell cycle

The cell cycle can be subdivided into two stages: interphase and mitosis. Genome replication occurs during the interphase, and its segregation to the daughter cells during the mitosis. The interphase includes G_1, S, and G_2 phases. Cells in G_0 are not actively cycling and have to be stimulated by growth factors in order to enter the cell cycle in G_1 [49]. Mitosis includes prophase, prometaphase, metaphase, anaphase, and telophase, and also cell division (cytokinesis), which overlaps the final stages of mitosis [50, 51]. In this chapter we will further analyze only the interphase.

DNA synthesis and doubling of the genome take place during the synthetic or S phase. This is preceded by a period or gap of variable duration called G_1 during which the cell is preparing for DNA synthesis, and is followed by a period known as the second gap or G_2, during which the cell prepares for mitosis [48, 52].

3.2.1.1. Cell cycle regulation

Cell proliferation is a process fundamental to development, growth, homeostasis, adaptation to disease, and neoplasia. For this reason, cell cycle events must be tightly regulated to ensure that they occur in the correct order with respect to each other and that they occur only once per cell cycle [53].

At least two types of cell cycle regulation mechanisms have been recognized: cell cycle checkpoints, which are surveillance mechanisms that monitor the order, integrity, and fidelity of the major events of the cell cycle [54], and a cascade of activation and deactivation of a series of proteins that relay a cell from one stage to the next [47].

3.2.1.1.1. Cell cycle checkpoints

Cell cycle checkpoints are a series of control systems enabling proliferation only in the presence of stimulatory signals (e.g. growth factors). They also arrest the cell cycle in response to DNA damage in order to provide time for DNA repair. After damage repair, progression through the cell cycle resumes. If the damage cannot be repaired, the cell is eliminated by apoptosis [55].

The primary checkpoint acts late in G_1. Once the cell has entered S phase, it is bound to continue through S, G_2, and M and thus produce two daughter cells. This checkpoint is sometimes known as the "point of no return" in the cell cycle with respect to S phase entry [56].

Additional checkpoints exist in S phase to activate DNA repair mechanisms when necessary. Furthermore, incomplete DNA replication or DNA damage triggers checkpoint pathways that block the G_2/M transition to ensure that cells have completely replicated their DNA and that it is intact before they enter mitosis [57].

Finally, the spindle assembly checkpoint acts during mitosis to maintain genome stability by delaying cell division (cytokinesis) until accurate chromosome segregation can be guaranteed [58].

3.2.1.1.2. Cyclin-dependent kinases (Cdks) regulation

The main families of proteins that play key roles in controlling cell cycle progression are the Cdks, the cyclins, the Cdk inhibitors (CKIs), and the tumor-suppressor gene products—the retinoblastoma protein (pRb) and p53 [59].

Progression of the cell through the cell cycle is mediated by sequential activation and inactivation of Cdks. The Cdks are a family of serine/threonine protein kinases that are activated at specific points of the cell cycle by the cyclins. Cdks activity can be counteracted by cell cycle inhibitory proteins, the CKIs [60, 61].

Activated Cdks induce downstream processes by phosphorylating selected proteins. pRb is a downstream target of Cdks-cyclins complexes [62]. Full pRb hyperphosphorylation releases pRb from E2F relieving repression of E2F target genes and allowing for activated E2F-dependent transcriptional induction and cell cycle progression [63].

p53 is stabilized in response to DNA damage, oncogenic stress, and various other stress conditions and activates transcription of a number of genes (including *p21*, *Mdm2*, and *Bax*) that induce cell cycle arrest or apoptosis. At the G_1/S checkpoint (see above), cell cycle arrest induced by DNA damage is p53-dependent [64].

3.2.2. Identification and measurement of cell proliferation

Assessment of cell proliferation is often of relevance in biomedical science, and a range of techniques have evolved to identify and quantify the process, generally by recognition and calculation of the number of cells in S or M phase [65].

A variety of markers have been used to determine cell cycle status and quantify cell proliferation, including the identification of mitotic figures, tritiated thymidine incorporation, bromodeoxyuridine incorporation, expression of proteins such as the proliferative cell nuclear antigen (PCNA), Ki-67, cyclins and Cdks, and the analysis of Cdks phosphorylation status [62, 66].

Of importance for this chapter are the immunohistochemical methods that detect proliferation-associated antigens. Ideally, such methods should be applicable to routinely processed tissues, they should be relatively inexpensive and the results easily quantified and interpreted [67]. The best known markers employed to recognize proliferating cells are Ki-67 and PCNA.

Cells express Ki-67 during G_1, S, G_2, and M phases, but not during the resting phase G_0. Its levels are low in the G_1 and S phases and rise to their peak level in M. Later in the M phase, a sharp decrease in Ki-67 levels occurs [68]. Ki-67 is required to maintain individual mitotic chromosomes dispersed in the cytoplasm after their release from the nuclear envelope, through a surfactant mechanism [69].

Ki-67 is widely used as a proliferation marker because it provides a rapid and relatively inexpensive method of measuring dividing cells [65, 70]. However, the short half-life of Ki-67 (1–1.5 h, regardless of the cell position in the cell cycle [71, 72]) makes its detection difficult. Furthermore, some healthy tissues can express low levels of Ki-67 [68].

PCNA was first shown to act as a cofactor/auxiliary protein for DNA polymerase δ, which is required for DNA synthesis during replication. However, besides DNA replication, PCNA functions have been associated with other cellular processes such as chromatin remodeling, DNA repair, sister-chromatid cohesion, and cell cycle control [73]. During DNA replication, presence of PCNA is necessary for synthesis of the leading strand. Levels of PCNA expression are therefore highest during S phase, with little to no expression during G_1 and intermediate levels in G_2 and M phases [62, 74].

PCNA detection has been widely used in immunohistochemical studies of cell proliferation. However, some authors claim that PCNA is not a reliable marker of this process because it is a pleiotropic protein involved in several aspects of cell control and not only in proliferation [66]. On the opposite, other authors affirm that PCNA is the most reliable and versatile of all markers used to analyze cell proliferation [75]. In the past, we have successfully used the immunohistochemical detection of PCNA in studies of cell turnover in lung [76].

4. Stem cells

4.1. Definition and classification

Stem cells are generally defined as clonogenic cells capable of both self-renewal and multilineage differentiation [77]. For a cell to be considered a stem cell, it must be capable of asymmetrical cell division, producing an exact multipotent replica cell, and an additional progeny cell than can perform a more specialized function [78].

Stem cells are classified according to their origin and developmental status in embryonic stem cells (ESC) and adult stem cells. Embryonic stem cells (ESC) can be derived from the inner cell mass of a blastocyst during gastrulation. They are totipotent cells giving rise to the germ line during development and virtually to all tissues of the organism [78, 79]. Adult stem cells (ASC) are tissue-resident stem cells that, based on their differentiation potency, can be classified as multipotent, oligopotent, or even unipotent [80]. In their tissue of residency, ASC function as lineage-committed progenitors to cells capable of more highly specialized tasks [78]. They are involved in tissue homeostasis and repair after wounding over the lifetime [79].

Among the tissues and organs harboring ASC, there are bone marrow, vascular walls, adipose tissues, skeletal muscles, heart, and brain, as well as epithelium of lung, liver, pancreas, digestive tract, skin, retina, breast, ovaries, prostate, and testis [81]. The bone marrow stem cell niche includes the hematopoietic stem cell population, which provides continuous renewal of blood cell lineages and the foundation of the immune system, and the mesenchymal stem cell population, responsible for osteogenic, adipogenic, and chondrogenic differentiation [82].

4.2. Mesenchymal stem cells (MSCs)

The minimal criteria for defining MSCs include: (a) remain plastic-adherent under standard culture conditions; (b) express CD73, CD90, and CD105, and lack expression of CD34, CD45, CD14 or CD11b, CD79a or CD19, and HLA-DR surface molecules; and (c) differentiate into osteoblasts, adipocytes, and chondrocytes *in vitro* [83, 84].

Originally isolated from bone marrow, MSCs have being isolated from other sites including spleen, thymus, muscle, adipose tissue, endometrium, placenta, umbilical cord, umbilical cord blood, peripheral blood, periosteum, periodontal ligament, dental pulp, synovium, synovial fluid, tendons, and cartilage [84, 85]. A perivascular location for MSCs has been suggested, correlating these cells with pericytes. This would explain why MSCs can be virtually isolated from all tissues [79, 86].

MSCs have demonstrated significant potential for clinical use due to their convenient isolation, their lack of significant immunogenicity permitting allogenic transplantation, their lack of ethical controversy, and their potential to differentiate into tissue-specific cell types [87]. MSCs may have therapeutic applications in several clinical disorders including myocardial infarction, diabetes, sepsis, hepatic failure, acute renal failure, several kinds of lung disease, as well as in spinal cord injuries, and bone and cartilage diseases [88, 89].

4.3. Nestin-positive MSCs

The human nestin protein consists of 1621 amino acids and displays a predicted molecular weight of 177.4 kDa. It is a class VI intermediate filament protein. Intermediate filaments represent, along with microtubules and actin filaments, one of the main components of cytoskeleton in animal cells [90].

Although nestin was first described as a marker of neural stem cells [91], its expression has also been shown in various prenatal and adult cells and tissues. Nestin-expressing cell types in embryonic and fetal tissues includes developing skeletal muscle cells, developing cardiomyocytes, endothelial cells of developing blood vessels, pancreatic epithelial progenitor cells, and hepatic oval cells. In adult, nestin expression has been found in, for example, satellite cells in dorsal root ganglia, retina, pancreatic stellate and endothelial cells, interstitial cells of Cajal, muscularis propria, Sertolli cells, and odontoblasts. Nestin has also been found to be expressed in injured tissues and in cancer cells [92].

In most of the studies, nestin has been detected by immunohistochemistry [92]. The principal advantage of immunohistochemistry over other techniques is that it enables the observation of processes in the context of intact tissue [93].

Normally, nestin becomes up-regulated in tissues during embryogenesis and down-regulated during maturation. During tissue injury in the adult, nestin is expressed in cells with progenitor cell properties. Furthermore, observational and interventional studies in animals and humans have shown that nestin may be an important marker for MSCs. These cells seem to act as a tissue reserve and to participate in tissue repair, regeneration, and growth [94, 95].

5. Cell renewal in lung cartilage

Cartilage grows by two methods: appositional growth and interstitial growth. In the former, chondroblasts in the perichondrium are transformed into chondrocytes. Interstitial growth result from mitotic division of pre-existing chondrocytes within the matrix. These two mechanisms occur early in life [96].

In the past, it has been believed that healthy adult chondrocytes maintain a stable resting phenotype and resist proliferation and differentiation throughout life [5]. Most cell types reach cell cycle arrest after a characteristic number of population doublings. The limit for human chondrocytes has been estimated at ~35 population doublings [4]. Their decreasing proliferative potential has been attributed to replicative senescence associated with erosion of telomere length [97].

We analyzed lung specimens from adult mice embedded in paraffin. Apoptosis was analyzed by TUNEL assay. PCNA and nestin were examined by immunohistochemistry. Apoptosis and PCNA were detected in lung chondrocytes. Serial section analysis demonstrated that cells in apoptosis were different from PCNA-positive cells, indicating that turnover was occurring. Chondrocytes were negative for nestin. However, nestin-positive cells were found in connective tissue associated with cartilage, in some specimens in close proximity of it and in perivascular cells. Thus, the findings of this work indicated that cell turnover in adult lung cartilage is possible, and that it may be mediated by nestin-positive cells [98].

In another related work, we found nestin-positive cells inside of lung cartilage and cells in division very close from them. This finding indicated that there exist nestin-positive MSCs in the adult that are able to differentiate into lung chondrocytes, perhaps to maintain homeostasis and/or repair damaged tissue [99].

For a long time it has been considered that cartilage contains a unique type of cell: the chondrocyte. However, nestin-positive MSCs has been found in cultured human adult lung cells, which underwent chondrogenic differentiation [100], and evidence from our investigations [98, 99] indicates that besides chondrocytes there exist nestin-positive MSCs in the adult lung cartilage.

The nestin-positive MSCs might be circulating in the blood stream or remain located in local blood vessels and be able to populate the cartilage when necessary, and/or might reside inside it. Other authors have shown that murine MSCs embolised within pulmonary blood vessels following systemic injection, and then transmigrated and differentiated into cartilage [101].

Finally, in another work, we found nestin-positive cells in perivascular areas and in connective tissue that were in close proximity of the bronchial airway epithelium. Nestin-positive cells were also found among the cells lining the airway epithelium, perhaps in order to participate in epithelial renewal [102]. Thus, stem cell reported in our works might be a pluripotent cell, which are able to generate several types of lung tissues. Other researchers presented evidence that a pluripotent stem cell exists in the lung that can generate lung-like tissue *in vitro* [103, 104].

6. Conclusion

Most of cells, tissues, and organs show continuous turnover. A delicate balance between cell proliferation and cell death ensures the maintenance of normal tissue morphology and function. Stem cells play essential roles in the growth, homeostasis and repair of many tissues. MSCs can give rise to multiple lineages including bone, adipose, and cartilage. The intermediate filament protein nestin was initially identified as a marker for neural stem cells, but its expression has also been detected in many types of cells, including MSCs.

It is generally believed that adult cartilage has no capacity to renewal. Taken together, our findings indicate that there exist nestin-positive MSCs in healthy adulthood that participates in the turnover of lung cartilage and in lung airway epithelium renewal. These findings may improve our knowledge about the biology of the cartilage and of the stem cells, and could provide new cell candidates for cartilage tissue engineering and for therapy for devastating pulmonary diseases.

Author details

Yareth Gopar-Cuevas, Alberto Niderhauser-García, Adriana Ancer-Arellano,
Ivett C. Miranda-Maldonado, María-de-Lourdes Chávez-Briones, Laura E. Rodríguez-Flores,
Marta Ortega-Martínez and Gilberto Jaramillo-Rangel*

*Address all correspondence to: gjaramillorangel@yahoo.com.mx

Department of Pathology, School of Medicine, Autonomous University of Nuevo Leon,
Monterrey, Mexico

References

[1] Eroschenko VP. Di Fiore's Atlas of Histology with Functional Correlations. 11th ed. Philadelphia: Lippincott Williams & Wilkins; 2008. 532 p

[2] Sandell LJ. Metabolism of chondrocytes in osteoarthritis: Why all this activity? Journal of Musculoskeletal & Neuronal Interactions. 2008;**8**:307

[3] Huber M, Trattnig S, Lintner F. Anatomy, biochemistry, and physiology of articular cartilage. Investigative Radiology. 2000;**35**:573-580

[4] Martin JA, Buckwalter JA. Roles of articular cartilage aging and chondrocyte senescence in the pathogenesis of osteoarthritis. Iowa Orthopedic Journal. 2001;**21**:1-7

[5] Umlauf D, Frank S, Pap T, Bertrand J. Cartilage biology, pathology, and repair. Cellular and Molecular Life Sciences. 2010;**67**:4197-4211

[6] Cooper GM, Hausman RE. The Cell: A Molecular Approach. 6th ed. Sunderland: Sinauer Associates; 2013. 832 p

[7] Eurell JA, Frappier BL. Dellmann's Textbook of Veterinary Histology. 6th ed. Ames: Blackwell Publishing; 2006. 420 p

[8] Muir H. The chondrocyte, architect of cartilage. Biomechanics, structure, function and molecular biology of cartilage matrix macromolecules. BioEssays. 1995;**17**:1039-1048

[9] Goldring MB, Marcu KB. Cartilage homeostasis in health and rheumatic diseases. Arthritis Research & Therapy. 2009;**11**:224

[10] Sadler TW. Langman's Medical Embryology. 12th ed. Philadelphia: Lippincott Williams & Wilkins; 2012. 384 p

[11] Naumann A, Dennis JE, Awadallah A, Carrino DA, Mansour JM, Kastenbauer E, Caplan AL. Immunochemical and mechanical characterization of cartilage subtypes in rabbit. The Journal of Histochemistry and Cytochemistry. 2002;**50**:1049-1058

[12] Cormack DH. Essential Histology. 2nd ed. Philadelphia: Lippincott Williams & Wilkins; 2001. 463 p

[13] Henrikson RC, Kaye GI, Mazurkiewicz JE. Respiratory system. In: Nieginski EA, editor. Histology. The National Medical Series for Independent Study. New York: Lippincott Williams & Wilkins; 1997. p. 311-321

[14] Dennison RD, Mathews PJ. Respiratory care. In: Mayer B, Kowalak JP, editors. Illustrated Manual of Nursing Practice. New York: Lippincott Williams & Wilkins; 2002. p. 237

[15] Ortega-Martínez M, Gopar-Cuevas Y, Cerda-Flores RM, Ancer-Arellano A, Chávez-Briones ML, De-la-Garza-González C, Rodríguez-Flores LE, Ancer-Rodríguez J, Jaramillo-Rangel G. Morphometric analysis of the bronchiolar arterioles through the normal aging process. In: Méndez-Vilas A, editor. Microscopy and Imaging Science:

Practical Approaches to Applied Research and Education. Badajoz: Formatex Research Center; 2017. p. 289

[16] Janssens JP, Pache JC, Nicod LP. Physiological changes in respiratory function associated with ageing. The European Respiratory Journal. 1999;**13**:197-205

[17] Shaffer TH, Wolfson MR, Panitch HB. Airway structure, function and development in health and disease. Paediatric Anaesthesia. 2004;**14**:3-14

[18] Sharma G, Goodwin J. Effect of aging on respiratory system physiology and immunology. Clinical Interventions in Aging. 2006;**1**:253-260

[19] Reid L. Visceral cartilage. Journal of Anatomy. 1976;**122**:349-355

[20] Medh RD, Thompson EB. Hormonal regulation of physiological cell turnover and apoptosis. Cell and Tissue Research. 2000;**301**:101-124

[21] Pellettieri J, Sánchez Alvarado A. Cell turnover and adult tissue homeostasis: From humans to planarians. Annual Review of Genetics. 2007;**41**:83-105

[22] Anti M, Armuzzi A, Gasbarrini A, Gasbarrini G. Importance of changes in epithelial cell turnover during *Helicobacter pylori* infection in gastric carcinogenesis. Gut. 1998;**43**:S27-S32

[23] Yun MH. Changes in regenerative capacity through lifespan. International Journal of Molecular Sciences. 2015;**16**:25392-25432

[24] Sousounis K, Baddour JA, Tsonis PA. Aging and regeneration in vertebrates. Current Topics in Developmental Biology. 2014;**108**:217-246

[25] Matute-Bello G, Martin TR. Science review: Apoptosis in acute lung injury. Critical Care. 2003;**7**:355-358

[26] Kerr JF, Wyllie AH, Currie AR. Apoptosis: A basic biological phenomenon with wide-ranging implications in tissue kinetics. British Journal of Cancer. 1972;**26**:239-257

[27] Reed JC, Tomaselli KJ. Drug discovery opportunities from apoptosis research. Current Opinion in Biotechnology. 2000;**11**:586-592

[28] Otsuki Y, Li Z, Shibata MA. Apoptotic detection methods-from morphology to gene. Progress in Histochemistry and Cytochemistry. 2003;**38**:275-339

[29] Elmore S. Apoptosis: A review of programmed cell death. Toxicologic Pathology. 2007;**35**:495-516

[30] Galluzzi L, Vitale I, Abrams JM, Alnemri ES, Baehrecke EH, Blagosklonny MV, Dawson TM, Dawson VL, El-Deiry WS, Fulda S, Gottlieb E, Green DR, Hengartner MO, Kepp O, Knight RA, Kumar S, Lipton SA, Lu X, Madeo F, Malorni W, Mehlen P, Nuñez G, Peter ME, Piacentini M, Rubinsztein DC, Shi Y, Simon HU, Vandenabeele P, White E, Yuan J, Zhivotovsky B, Melino G, Kroemer G. Molecular definitions of cell death subroutines: Recommendations of the Nomenclature Committee on Cell Death 2012. Cell Death and Differentiation. 2012;**19**:107-120

[31] Kroemer G, Galluzzi L, Vandenabeele P, Abrams J, Alnemri ES, Baehrecke EH, Blagosklonny MV, El-Deiry WS, Golstein P, Green DR, Hengartner M, Knight RA, Kumar S, Lipton SA, Malorni W, Nuñez G, Peter ME, Tschopp J, Yuan J, Piacentini M, Zhivotovsky B, Melino G. Nomenclature Committee on Cell Death 2009. Classification of cell death: Recommendations of the Nomenclature Committee on Cell Death 2009. Cell Death and Differentiation. 2009;**16**:3-11

[32] Lowe SW, Lin AW. Apoptosis in cancer. Carcinogenesis. 2000;**21**:485-495

[33] Igney FH, Krammer PH. Death and anti-death: Tumour resistance to apoptosis. Nature Reviews. Cancer. 2002;**2**:277-288

[34] Martinvalet D, Zhu P, Lieberman J, Granzyme A. Induces caspase-independent mitochondrial damage, a required first step for apoptosis. Immunity. 2005;**22**:355-370

[35] Martinez MM, Reif RD, Pappas D. Detection of apoptosis: A review of conventional and novel techniques. Analytical Methods. 2010;**2**:996-1004

[36] Watanabe M, Hitomi M, van der Wee K, Rothenberg F, Fisher SA, Zucker R, Svoboda KK, Goldsmith EC, Heiskanen KM, Nieminen AL. The pros and cons of apoptosis assays in the study of cells, tissues, and organs. Microscopy and Microanalysis. 2002;**8**:375-391

[37] Soini Y, Pääkkö P, Lehto VP. Histopathological evaluation of apoptosis in cancer. The American Journal of Pathology. 1998;**153**:1041-1053

[38] Bonner-Weir S. Beta-cell turnover: Its assessment and implications. Diabetes. 2001;**50**: S20-S24

[39] Otsuki Y. Various methods of apoptosis detection. Acta Histochemica et Cytochemica. 2000;**33**:235-241

[40] Gavrieli Y, Sherman Y, Ben-Sasson SA. Identification of programmed cell death in situ via specific labeling of nuclear DNA fragmentation. The Journal of Cell Biology. 1992;**119**:493-501

[41] Kressel M, Groscurth P. Distinction of apoptotic and necrotic cell death by in situ labelling of fragmented DNA. Cell and Tissue Research. 1994;**278**:549-556

[42] Csizmadia E, Csizmadia V. Detection of apoptosis in tissue sections. Methods in Molecular Biology. 2009;**559**:49-63

[43] Migheli A, Attanasio A, Schiffer D. Ultrastructural detection of DNA strand breaks in apoptotic neural cells by in situ end-labelling techniques. The Journal of Pathology. 1995;**176**:27-35

[44] Ansari B, Coates PJ, Greenstein BD, Hall PA. Situ end-labelling detects DNA strand breaks in apoptosis and other physiological and pathological states. The Journal of Pathology. 1993;**170**:1-8

[45] Grasl-Kraupp B, Ruttkay-Nedecky B, Koudelka H, Bukowska K, Bursch W, Schulte-Hermann R. In situ detection of fragmented DNA (TUNEL assay) fails to discriminate among apoptosis, necrosis, and autolytic cell death: A cautionary note. Hepatology. 1995;**21**:1465-1468

[46] Berridge MJ. Cell cycle and proliferation. Cell Signalling Biology. 2014;**6**:1-45

[47] Collins K, Jacks T, Pavletich NP. The cell cycle and cancer. Proceedings of the National Academy of Sciences of the United States of America. 1997;**94**:2776-2778

[48] Vermeulen K, Van Bockstaele DR, Berneman ZN. The cell cycle: A review of regulation, deregulation and therapeutic targets in cancer. Cell Proliferation. 2003;**36**:131-149

[49] Norbury C, Nurse P. Animal cell cycles and their control. Annual Review of Biochemistry. 1992;**61**:441-470

[50] Strachan T, Read A. Human Molecular Genetics. 4th ed. New York: Garland Science; 2011. 782 p

[51] Patel S, Tomar RS, Gajera H, Golakiya BA, Parakhia MV. Handbook of Life Sciences. New Delhi: New India Publishing Agency; 2010. 792 p

[52] Hall PA, Levison DA. Review: Assessment of cell proliferation in histological material. Journal of Clinical Pathology. 1990;**43**:184-192

[53] Humphrey T, Pearce A. Cell cycle molecules and mechanisms of the budding and fission yeasts. In: Humphrey T, Brooks G, editors. Cell Cycle Control: Mechanisms and Protocols. Totowa: Humana Press; 2005. p. 3

[54] Barnum KJ, O'Connell MJ. Cell cycle regulation by checkpoints. Methods in Molecular Biology. 2014;**1170**:29-40

[55] Pucci B, Kasten M, Giordano A. Cell cycle and apoptosis. Neoplasia. 2000;**2**:291-299

[56] Zetterberg A, Larsson O, Wiman KG. What is the restriction point? Current Opinion in Cell Biology. 1995;**7**:835-842

[57] Harper JV, Brooks G. The mammalian cell cycle: An overview. In: Humphrey T, Brooks G, editors. Cell Cycle Control: Mechanisms and Protocols. Totowa: Humana Press; 2005. p. 114

[58] Lara-Gonzalez P, Westhorpe FG, Taylor SS. The spindle assembly checkpoint. Current Biology. 2012;**22**:R966-R980

[59] Golias CH, Charalabopoulos A, Charalabopoulos K. Cell proliferation and cell cycle control: A mini review. International Journal of Clinical Practice. 2004;**58**:1134-1141

[60] Pines J. Cyclins and cyclin-dependent kinases: Theme and variations. Advances in Cancer Research. 1995;**66**:181-212

[61] Morgan DO. Principles of CDK regulation. Nature. 1995;**374**:131-134

[62] Schafer KA. The cell cycle: A review. Veterinary Pathology. 1998;**35**:461-478

[63] Matson JP, Cook JG. Cell cycle proliferation decisions: The impact of single cell analyses. The FEBS Journal. 2017;**284**:362-375

[64] Wiman KG, Zhivotovsky B. Understanding cell cycle and cell death regulation provides novel weapons against human diseases. Journal of Internal Medicine. 2017;**281**:483-495

[65] Boulton RA, Hodgson HJ. Assessing cell proliferation: A methodological review. Clinical Science (London, England). 1995;**88**:119-130

[66] Goodlad RA. Quantification of epithelial cell proliferation, cell dynamics, and cell kinetics in vivo. Wiley Interdisciplinary Reviews: Developmental Biology. 2017:e274

[67] Alison MR. Assessing cellular proliferation: what's worth measuring? Human & Experimental Toxicology. 1995;**14**:935-944

[68] Yerushalmi R, Woods R, Ravdin PM, Hayes MM, Gelmon KA. Ki67 in breast cancer: Prognostic and predictive potential. The Lancet Oncology. 2010;**11**:174-183

[69] Cuylen S, Blaukopf C, Politi AZ, Müller-Reichert T, Neumann B, Poser I, Ellenberg J, Hyman AA, Gerlich DW. Ki-67 acts as a biological surfactant to disperse mitotic chromosomes. Nature. 2016;**535**:308-312

[70] Whitfield ML, George LK, Grant GD, Perou CM. Common markers of proliferation. Nature Reviews. Cancer. 2006;**6**:99-106

[71] Bruno S, Darzynkiewicz Z. Cell cycle dependent expression and stability of the nuclear protein detected by Ki-67 antibody in HL-60 cells. Cell Proliferation. 1992;**25**:31-40

[72] Heidebrecht HJ, Buck F, Haas K, Wacker HH, Parwaresch R. Monoclonal antibodies Ki-S3 and Ki-S5 yield new data on the "Ki-67" proteins. Cell Proliferation. 1996;**29**:413-425

[73] Strzalka W, Ziemienowicz A. Proliferating cell nuclear antigen (PCNA): A key factor in DNA replication and cell cycle regulation. Annals of Botany. 2011;**107**:1127-1140

[74] Emmett SR, Dove B, Mahoney L, Wurm T, Hiscox JA. The cell cycle and virus infection. In: Humphrey T, Brooks G, editors. Cell Cycle Control: Mechanisms and Protocols. Totowa: Humana Press; 2005. p. 208

[75] Iatropoulos MJ, Williams GM. Proliferation markers. Experimental and Toxicologic Pathology. 1996;**48**:175-181

[76] Ortega-Martínez M, Rodríguez-Flores LE, Ancer-Arellano A, Cerda-Flores RM, De-la-Garza-González C, Ancer-Rodríguez J, Jaramillo-Rangel G. Analysis of cell turnover in the bronchiolar epithelium through the normal aging process. Lung. 2016;**194**:581-587

[77] Weissman IL. Stem cells: Units of development, units of regeneration, and units in evolution. Cell. 2000;**100**:157-168

[78] Sylvester KG, Longaker MT. Stem cells: Review and update. Archives of Surgery. 2004;**139**:93-99

[79] Deregibus MC, Tetta C, Camussi G. The dynamic stem cell microenvironment is orchestrated by microvesicle-mediated transfer of genetic information. Histology and Histopathology. 2010;**25**:397-404

[80] Lazzeri E, Peired A, Ballerini L, Lasagni L. Adult stem cells in tissue homeostasis and disease. In: Najman S, editor. Current Frontiers and Perspectives in Cell Biology. InTech: Rijeka; 2012. p. 359

[81] Mimeault M, Batra SK. Recent progress on tissue-resident adult stem cell biology and their therapeutic implications. Stem Cell Reviews. 2008;**4**:27-49

[82] Green DE, Adler BJ, Chan ME, Rubin CT. Devastation of adult stem cell pools by irradiation precedes collapse of trabecular bone quality and quantity. Journal of Bone and Mineral Research. 2012;**27**:749-759

[83] Dominici M, Le Blanc K, Mueller I, Slaper-Cortenbach I, Marini F, Krause D, Deans R, Keating A, Prockop DJ, Horwitz E. Minimal criteria for defining multipotent mesenchymal stromal cells. The International Society for Cellular Therapy position statement. Cytotherapy. 2006;**8**:315-317

[84] Lv FJ, Tuan RS, Cheung KM, Leung VY. Concise review: The surface markers and identity of human mesenchymal stem cells. Stem Cells. 2014;**32**:1408-1419

[85] Pountos I, Giannoudis PV. Biology of mesenchymal stem cells. Injury. 2005;**36**:S8-S12

[86] Crisan M, Yap S, Casteilla L, Chen CW, Corselli M, Park TS, Andriolo G, Sun B, Zheng B, Zhang L, Norotte C, Teng PN, Traas J, Schugar R, Deasy BM, Badylak S, Buhring HJ, Giacobino JP, Lazzari L, Huard J, Péault B. A perivascular origin for mesenchymal stem cells in multiple human organs. Cell Stem Cell. 2008;**3**:301-313

[87] Karp JM, Leng Teo GS. Mesenchymal stem cell homing: The devil is in the details. Cell Stem Cell. 2009;**4**:206-216

[88] Barry FP, Murphy JM. Mesenchymal stem cells: Clinical applications and biological characterization. The International Journal of Biochemistry & Cell Biology. 2004;**36**:568-584

[89] Lee JW, Fang X, Krasnodembskaya A, Howard JP, Matthay MA. Concise review: Mesenchymal stem cells for acute lung injury: Role of paracrine soluble factors. Stem Cells. 2011;**29**:913-919

[90] Neradil J, Veselska R. Nestin as a marker of cancer stem cells. Cancer Science. 2015;**106**:803-811

[91] Lendahl U, Zimmerman LB, McKay RD. CNS stem cells express a new class of intermediate filament protein. Cell. 1990;**60**:585-595

[92] Wiese C, Rolletschek A, Kania G, Blyszczuk P, Tarasov KV, Tarasova Y, Wersto RP, Boheler KR, Wobus AM. Nestin expression—A property of multi-lineage progenitor cells? Cellular and Molecular Life Sciences. 2004;**61**:2510-2522

[93] Webster JD, Miller MA, DuSold D, Ramos-Vara J. Effects of prolonged formalin fixation on the immunohistochemical detection of infectious agents in formalin-fixed, paraffin-embedded tissues. Veterinary Pathology. 2010;**47**:529-535

[94] Tampaki EC, Nakopoulou L, Tampakis A, Kontzoglou K, Weber WP, Kouraklis G. Nestin involvement in tissue injury and cancer—A potential tumor marker? Cellular Oncology (Dordrecht). 2014;**37**:305-315

[95] Xie L, Zeng X, Hu J, Chen Q. Characterization of nestin, a selective marker for bone marrow derived mesenchymal stem cells. Stem Cells International. 2015;**2015**:762098

[96] Martini FH. Anatomy & Physiology. 1st ed. Jurong: Pearson Education, Inc.; 2005. 845 p

[97] Martin JA, Brown TD, Heiner AD, Buckwalter JA. Chondrocyte senescence, joint loading and osteoarthritis. Clinical Orthopaedics and Related Research. 2004;**427**(Suppl):S96-103

[98] Ortega-Martínez M, Romero-Núñez E, Niderhauser-García A, De-la-Garza-González C, Ancer-Rodríguez J, Jaramillo-Rangel G. Evidence of chondrocyte turnover in lung cartilage, with the probable participation of nestin-positive cells. Cell Biology International. 2013;**37**:239-241

[99] Ortega-Martínez M, De-la-Garza-González C, Ancer-Rodríguez J, Jaramillo-Rangel G. Nestin-positive stem cells participate in chondrocyte renewal in healthy adult lung cartilage. International Journal of Morphology. 2014;**32**:151-153

[100] Sabatini F, Petecchia L, Tavian M, Jodon de Villeroché V, Rossi GA, Brouty-Boyé D. Human bronchial fibroblasts exhibit a mesenchymal stem cell phenotype and multilineage differentiating potentialities. Laboratory Investigation. 2005;**85**:962-971

[101] Aguilar S, Nye E, Chan J, Loebinger M, Spencer-Dene B, Fisk N, Stamp G, Bonnet D, Janes SM. Murine but not human mesenchymal stem cells generate osteosarcoma-like lesions in the lung. Stem Cells. 2007;**25**:1586-1594

[102] Ortega-Martínez M, Rodríguez-Flores LE, De-la-Garza-González C, Ancer-Rodríguez J, Jaramillo-Rangel G. Detection of a novel stem cell probably involved in normal turnover of the lung airway epithelium. Journal of Cellular and Molecular Medicine. 2015;**19**:2679-2681

[103] Cortiella J, Kojima K, Bonassar LJ, Hendricks G, Vacanti CA, Vacanti MP. Tissue engineered lung. Tissue Engineering. 2000;**6**:661

[104] Vacanti MP, Roy A, Cortiella J, Bonassar L, Vacanti CA. Identification and initial characterization of spore-like cells in adult mammals. Journal of Cellular Biochemistry. 2001;**80**:455-460

Chapter 3

Alternative Therapeutic Approach for Cartilage Repair

Marina Cristina Akuri, Mariana Ricci Barion,
Sandra Maria Barbalho and Élen Landgraf Guiguer

Additional information is available at the end of the chapter

http://dx.doi.org/10.5772/intechopen.72478

Abstract

The cartilage is a flexible tissue, which supports the adjacent soft tissues. The damages that cause degenerative articular diseases are marked by the increase of cytokines such as tumor necrosis factor-α (TNF-α), IL-1β, IL-6, IL-18, and IL-17, which cause intense inflammatory process and release of metalloproteinases and disintegrin enzymes that lead to cartilage degradation. The *Curcuma longa* possesses bioactive compounds designated as curcuminoids that display therapeutic potential in several pathologies. Curcumin is one of these compounds that may exhibit anti-inflammatory, antioxidant, antiviral, antibacterial, and antitumor effects. It may promote decrease of IL-1β, IL-6, IL-8, TNF-α, COX-2, and reactive oxygen species. Furthermore, curcumin inhibits the activity of several kinases related to the degradation of the cartilage, including tyrosine kinase, p21-activated kinase, mitogen-activated protein kinase, protein kinase C, the activator protein 1 pathway, and NF-κB leading to the suppression of the production of metalloproteinases and inflammatory cytokines. Curcumin has also been related to the stimulation of the production of type II collagen and glycosaminoglycan by chondrocytes. Studies have shown that this compound may alleviate joint pain and crepitation, reduce the use of other drugs for pain relief, stimulate the production of type II collagen and glycosaminoglycan resulting in a protective and anti-inflammatory action of cartilage and bones, and improve the quality of life of the patients.

Keywords: cartilage, inflammation, *Curcuma longa*, curcumin

1. Introduction

The articular cartilage is a flexible tissue, which supports the adjacent soft tissues and possesses the extracellular matrix (ECM), collagen, chondrocyte, proteoglycans, and water [1]. This tissue is alymphatic, avascular, and aneural, and for these reasons, when a severe damage occurs, the self-repair is a highly difficult process [2–4].

The damages that cause degenerative articular diseases are marked by the increase of cytokines that cause intense inflammatory process and enzymes that cause cartilage degradation [5, 6].

The osteoarthritis (OA) is an example of a progressive degenerative disease characterized by a chronic inflammatory process, joint pain, and loss of function and injury of adjacent tissue. The great destruction of the articular cartilage is the main characteristic of this disease [7–9], and therefore, it is used, in this chapter, as a prototype of cartilage destruction and regeneration.

Drugs such as nonsteroidal anti-inflammatory drugs (NSAIDS) and acetaminophen are the therapeutic approaches for the pharmacological treatment of degenerative diseases. However, this kind of medications is associated with gastrointestinal, cardiovascular, and renal adverse effects and do not effectively inhibit the disease progression and destruction of cartilage [7, 8, 10, 11]. Furthermore, corticosteroids, another therapeutic option due to their potent anti-inflammatory action and ability to reduce symptoms, should also not be used for an extended period because they can lead to a more rapid progression of OA [12].

The development of therapeutic alternatives that do not cause adverse effects and inhibit the progression of the disease is urgent and, therefore, has been widely studied. *Curcuma longa*, herbal medicine, has been shown to be one of these possible alternatives because it presents significant benefits in degenerative diseases such as OA, and this plant may play a crucial role in the reduction of the inflammatory pathways [13].

2. Physiopathology and cartilage destruction

In healthy cartilage, chondrocytes can form ECM components and enzymes that degrade cartilage in equilibrium. Although the pathophysiology of OA and its triggers have not yet been fully elucidated, it is known that inflammation, joint destruction, synovitis, and osteoclastogenesis are involved [9].

In OA, there is an increase in the enzymes involved in the cartilage degradation such as disintegrin and metalloproteinase (MMP) with a thrombospondin motif (ADAMTS). This enzymatic increase occurs due to stimulation by interleukins (IL) and inflammatory mediators such as tumor necrosis factor-α (TNF-α), IL-1β, IL-6, IL-18, and IL-17 [6, 14].

The most widely studied and related to the destruction of cartilage are ADAMTS4 and ADAMTS5 that are released after stimulation of inflammatory cytokines such as IL-1β. They are aggrecanases and aggrecan aggregation of proteoglycans and one of the components of ECM. After degradation of aggrecans by these enzymes, MMP-3 acts in synergism in the degradation of proteoglycans [8, 15–17].

MMPs are enzymes implicated primarily in the destruction of type II collagen and therefore play a fundamental role in the destruction of cartilage. MMP-1, MMP-3, MMP-9, and MMP-13 are the most involved enzymes in this process, and the last one is not found in adult cartilage without OA. Fragments from cleavage of collagen type 2 by MMPs amplify the destruction of ECM and amplify the release of more MMPs [5–9].

http://dx.doi.org/10.5772/intechopen.72478

Activated synoviocytes also produce inflammatory cytokines such as IL-1, IL-6, and TNF-α, which act by amplifying the inflammatory process and cartilaginous destruction. There is increased release of reactive oxygen species (ROS), mainly nitric oxide (NO), peroxynitrite ($ONOO^{-}$), and superoxide anion radicals ($O2^{-}$). Other inflammatory mediators such as cyclo-oxygenase-2 (COX-2), produced by synovial monocytes, and prostaglandins 2 (PGE2) are also involved in the pathophysiology of the disease [6, 15].

The nuclear factor-kappa B (NF-κβ) pathway is responsible for the production of various cytokines and induction of inflammation. When stimulated by interleukins IL-1β and TNF-α, there is activation of I kappa beta kinase (IKK), which promotes the phosphorylation of IKB-α. Thus, IKB-α is degraded by ubiquitination, and the dimers compounded by p50 and p65 reach the nucleus and can stimulate the expression of more than 400 genes, of which some are pro-inflammatory and pro-apoptotic genes [8, 18]. Therefore, there is production of various interleukins, including IL-1, IL-6, IL-8, and IL-10 [8, 15, 16].

Besides the destruction of cartilage in OA, an intense process of bone resorption occurs. This process is a result of osteoclast activation known as osteoclastogenesis [19]. The receptor activator NF-kappa ligand (RANKL) is produced by some cells as the osteoblast and has an affinity for RANK, which is present in the membrane of osteoclast precursor cells [20]. When RANKL binds to RANK, a phosphorylation process occurs, culminating in the activation of NF-kB [5, 21, 22]. The osteoprotegerin also has an affinity for RANK, thus competing with RANKL, inducing apoptosis of mature osteoclasts [5]. In the OA, the increase of RANKL and the decrease of OPG are observed [23, 24]. **Figure 1** summarizes the inflammatory process in the cartilage.

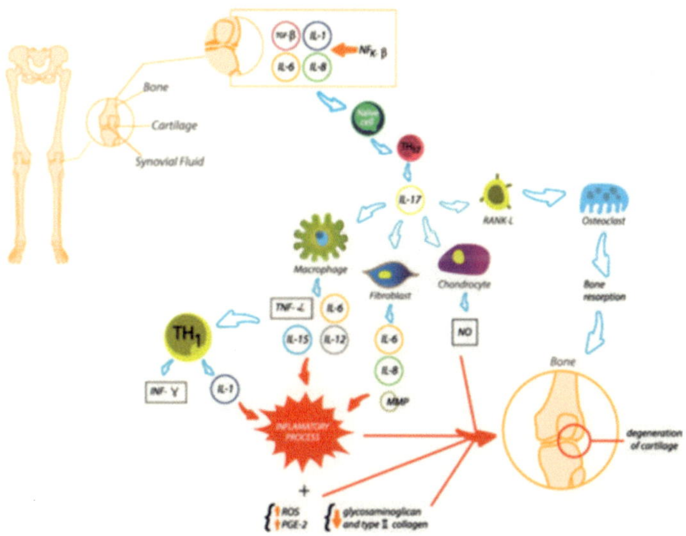

Figure 1. The activation of NF-kβ is related to the release of TGFβ, IL-1, IL-6, and IL-8 and further activation of TH17 that leads to the stimulation of several cells and expression of other inflammatory cytokines and metalloproteinases (MMP), and further development of features characteristic of inflammation and degradation of cartilage and bone. NF-kβ: nuclear factor kβ; IL: interleukin; TH17: T-helper 17; MMP: matrix metalloproteinase; RANKL: receptor activator of nuclear factor κB ligand; TH17: type 17 T-helper; TNF-α: tumor necrosis factor-α; TGF-β, transforming growth factor-β; ROS: reactive oxygen species; and NO: nitric oxide.

3. *Curcuma longa*

C. longa, or turmeric or saffron, is native to Asia and India and belongs to the Zingiberaceae family, and its rhizome has been used as a seasoning and in the traditional medicine since ancient times [18, 25].

The bioactive compounds derived from turmeric are called curcuminoids and have shown therapeutic potential in various pathologies. The three most important compounds originated from this rhizome are curcumin (diferuloylmethane), bisdemethoxycurcumin, and demethoxycurcumin, which are present, respectively, in concentrations of 77, 17, and 3%. Curcumin gives the typical yellowish coloration of the rhizome, and this part of the plant is the most widely studied [18, 26, 27].

Several studies have been conducted in order to show their actions *in vitro* and *in vivo*. Curcumin acts with different mechanisms and in different cell types and pathways. It shows anti-inflammatory, antioxidant, antiviral, antibacterial, and antitumor effects. Its therapeutic potential covers diseases such as cancer, Alzheimer's disease, osteoporosis, inflammatory bowel disease, depression, arthritis, diabetes, vitiligo, endometriosis, and several others. **Figure 2** shows some effects of curcumin [5, 28–30].

Studies on the action of curcumin and its analogs show that it can act directly or indirectly in the decrease of the formation of inflammatory molecules and pro-inflammatory transcription factors. Under its action, there is a reduction of IL-1β, IL-6, IL-8, TNF-α, NF-kB, COX-2, and reactive oxygen species (ROS). Apart from that, curcumin has been shown to inhibit the activity of several kinases related to the degradation of the cartilage, including a tyrosine

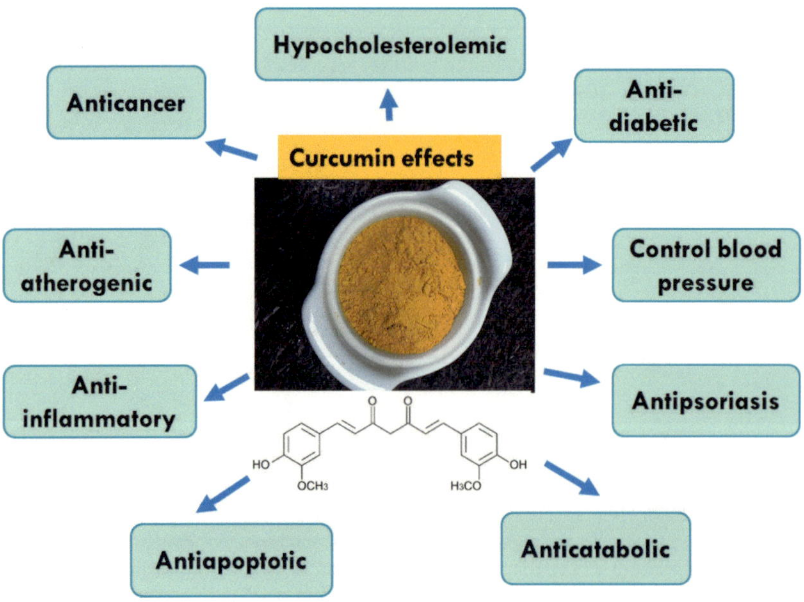

Figure 2. Some benefits of curcumin on human health.

kinase, p21-activated kinase 1 (PAK1), mitogen-activated protein kinase (MAPK), and protein kinase C (PKC). **Figure 3** shows the process of cartilage inflammation and the effects of curcumin in the healing process [29, 31, 32].

Many studies have shown that curcumin has potent effects on the induction of apoptosis and decreased tumor cell proliferation and may promote the inhibition of important angiogenesis regulators, signal transducers and activators of transcription 3 (STAT3), and vascular endothelial growth factor (VEGF). Besides, it downregulates the expression of differentiated embryo-chondrocyte expressed gene 1 (DEC1) and hypoxia-inducible factor-1-α (HIF-1α) [33–36].

Furthermore, several authors have shown that the supplementation with curcumin may bring a plethora of benefits in the treatment and prevention of the osteopenia [37]. This compound has been demonstrated to be able to avert the suppression of osteoblasts proliferation and to enhance the index of osteoprotegerin and RANKL, which indicates osteoblastogenesis [38].

As mentioned earlier, the actions of curcumin vary from potent anti-inflammatory and anti-apoptotic to antioxidant [39]. The wide variety of sites of actions and consequently decrease in the inflammation markers make this compound and its analogs extremely promising in chronic inflammatory diseases such as OA [5, 28]. Also, this herbal medicine inhibits the phosphorylation of IKB-α and thereby reduces cartilage degradation, as shown in **Figure 4**.

Conventional OA therapies are restricted to the reduction of symptoms in patients, but they do not decrease the degradation of cartilage and, consequently, do not alter the progression of the disease. For these reasons, the need for new therapies is striking, and curcumin and its analogs have become extremely promising in this context [13, 28].

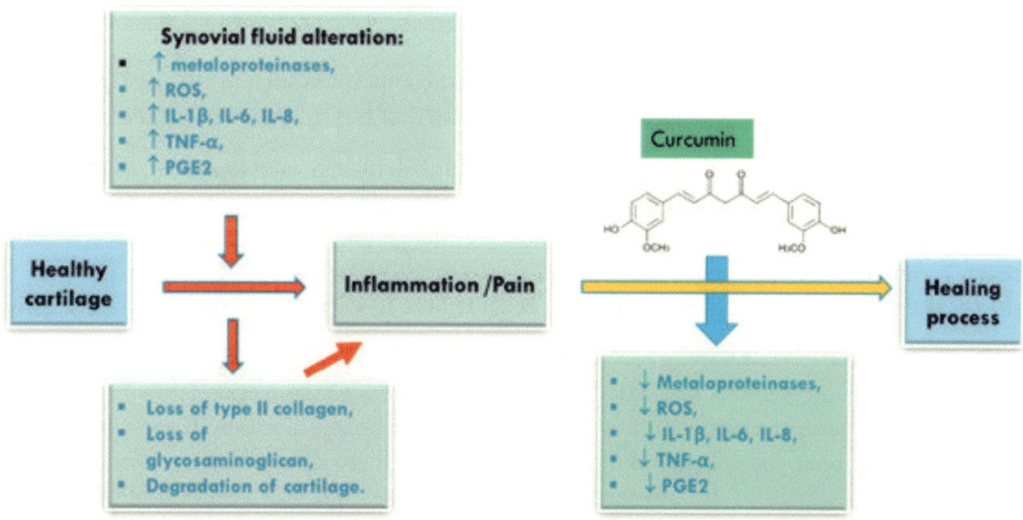

Figure 3. The inflammation of the cartilage may occur due to several processes such as an increase in the expression of enzymes, increase in the formation of ROS, and release of cytokines. The consequence is the loss of type II collagen and glycosaminoglycan resulting in the degradation of the cartilage. Curcumin interferes in this scenario and may help in the healing process. ROS: reactive oxygen species; IL: interleukin; TNF-α: tumor necrosis factor-α; PGE2: prostaglandin E2.

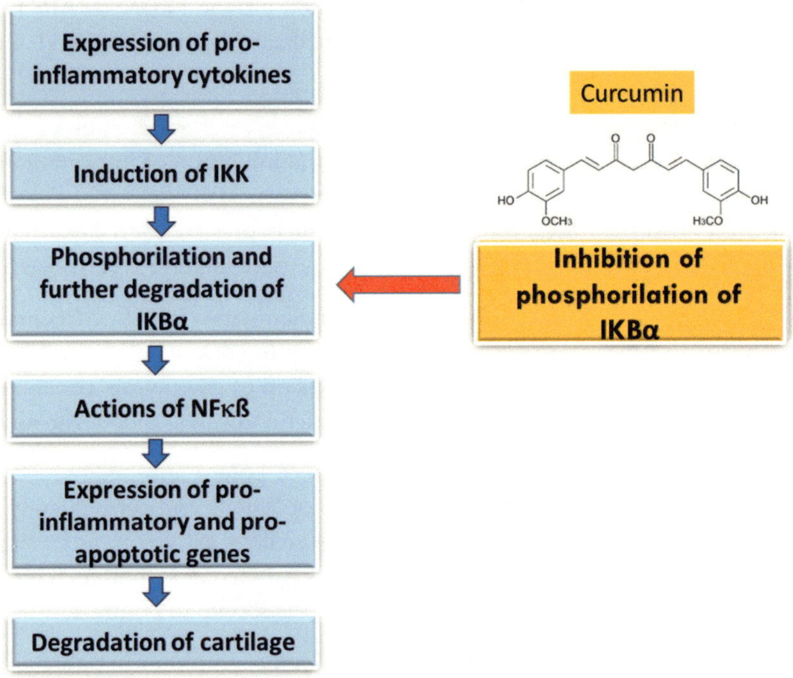

Figure 4. Effects of curcumin on the inhibition of the process involving the degradation of cartilage. IKK: I kappa B kinase; IKBα: inhibitor of kappa B; NF-κβ: nuclear factor κβ.

4. The potential effects of *Curcuma longa*

In articular cartilage, the ECM is composed of different compounds such as collagen, proteoglycans (glycosaminoglycan and proteins) mainly aggrecan and non-collagenous proteins [28, 40].

In the intra-articular space, there is the synovial fluid, which is enveloped by the synovial membrane and is also responsible for the nutrition of articular cartilage cells. The main cells in the synovial membrane are synoviocytes that have phagocytic functions and are responsible for the production of synovial fluid [41] and contribute to the inflammatory process when it releases several cytokines and proteases which contribute to joint destruction [42, 43].

Some degenerative diseases are involved with synovial inflammation and destruction of ECM of articular cartilage [44, 45]. According to the World Health Organization (WHO), musculoskeletal or rheumatic conditions consist over 150 syndromes and diseases. These ailments are liable for chronic pain, disability, and dysfunction. Among these diseases, rheumatoid arthritis, osteoarthritis, spinal disorders, and severe limb trauma deserve special mention because of the greatest impact on society such as healthcare expenditures. In developed countries, OA is one of the most disabling diseases [45].

OA is a degenerative disorder involving synovial inflammation and destruction of ECM leading to several symptoms such as pain, disability, and significant morbidity, requiring many medications that, in most cases, do not show effective actions resulting in the damage of the synovial tissue. For these reasons, new pharmacotherapies and therapies for this illness are essential [28].

As pointed earlier, curcumin may act in many different locals of inflammation resulting, directly or indirectly, in the reduction of the production of inflammatory mediators and interleukins, resulting in less destruction of cartilage. Besides that, patients treated with curcumin have decreased C-reactive protein, a marker of inflammation [46, 47].

Moreover, curcumin has been shown to inhibit the activator protein 1 (AP-1) pathway and NF-κB leading to the suppression of the production of MMP-3, MMP-9, and MMP-13 [15, 19]. Zhang et al. [9] demonstrated in a mouse model that the production of MMP-1, MMP-3, MMP-13, IL-1β, TNF-α, and ADAMTS5 was decreased when the animals were treated with curcumin. They also showed an increase in the expression of the chondroprotective gene CITED 2 (Cbp/P300 interacting transactivator with Glu/Asp rich carboxy terminal domain 2), which seems to be involved in the suppression of NF-κB activity [9, 19]. Curcumin has also been related to the stimulation of the production of type II collagen and glycosaminoglycan by chondrocytes [5].

Curcumin inhibits the activation of I kappa B kinase (IKK) in chondrocytes, osteoblasts, and synovial cells [15, 48]. By inhibiting the phosphorylation of this kinase, curcumin prevents the activation of NF-kB. Consequently, it inhibits the expression of pro-apoptotic genes in chondrocytes (caspase-3) and the formation of inflammatory mediators [18]. Thus, it is responsible for the downregulation of lipoxygenases, COX-2, phospholipase A2, prostaglandin E2 (PGE2), IL-1β, IL-6, and IL-8 [19, 30]. Wherefore, curcumin blocks the signaling by NF-kB, leading to the inhibition of this factor resulting in the decrease of the degradation of collagen. This pathway is induced by the activation of the chondrocytes stimulated by IL-1 [15, 16].

Curcumin inhibits TNF-α, which is associated with increased cartilage reabsorption. This cytokine associated with IL-6 and IL-1 inhibits the proteoglycan synthesis [5, 49, 50].

Studies have shown that compounds from *Curcuma* sp. can alleviate joint pain and crepitation, which lead to improved scores on WOMAC (Western Ontario and McMaster Universities Osteoarthritis Index), improve function, reduce the use of other drugs for pain relief, and is as effective as the use of ibuprofen [51–59].

Therefore, curcumin acts on the NF-kB system, in addition to the stimulation of the production of type II collagen and glycosaminoglycan resulting in a protective and anti-inflammatory action of cartilage and bones, reducing pain and improving the quality of life of patients with degenerative diseases [18].

5. Disadvantages of *Curcuma longa*

The major problem of curcumin is that it is extremely hydrophobic and thus has low oral bioavailability, thus decreasing their beneficial effects. Another problem is the rapid metabolism of curcuminoids considering the extensive biotransformation and consequent reduction in the plasmatic levels [25, 60].

Some techniques, such as nanoparticles, phospholipid complexes, and liposomes, have been used as drug delivery systems to improve the bioavailability of these substances [61, 62]. Some compounds, such as folic acid, piperine, phosphatidylcholine, galactose, and the complex arginine-glycine-aspartic acid, are also used to improve this bioavailability and effects. Green tea and collagen associated with curcumin extracts may also enhance its effects [8, 30, 63, 64].

6. Conclusions

The curcumin has been used as an alternative therapy in the control of cartilage healing once it may interfere with the inflammatory pathways reducing the release of pro-inflammatory cytokines. Nevertheless, the use of curcumin and its analogs need to be more extensively studied and tested to determine the bioavailability, the therapeutic properties, adequate delivery formulations, doses, and possible risks of use.

Author details

Marina Cristina Akuri[1], Mariana Ricci Barion[1], Sandra Maria Barbalho[1,2]* and Élen Landgraf Guiguer[1,2]

*Address all correspondence to: smbarbalho@gmail.com

1 Medical School of Marília, UNIMAR, Marília, São Paulo, Brazil

2 Food Technology School, Marília, São Paulo, Brazil

References

[1] Fernández-Torres J, Martínez-Nava GA, Gutiérrez-Ruíz MC, Gómez-Quiroz LE, Gutiérrez M. Role of HIF-1α signaling pathway in osteoarthritis: A systematic review. Revista Brasileira de Reumatologia English Edition. 2017;**57**:162-173. DOI: 10.1016/j.rbre.2016.07.008

[2] Zhou F, Zhang X, Cai D, Li J, Mu Q, Zhang W, Zhu S, Jiang Y, Shen W, Zhang S, Ouyang HW. Silk fibroin-chondroitin sulfate scaffold with immuno-inhibition property for articular cartilage repair. Acta Biomaterialia. 2017;**63**:64-75. DOI: 10.1016/j.actbio.2017.09.005 pii: S1742-7061(17)30569-X

[3] Ramezanifard R, Kabiri M, Hanaee Ahvaz H. Effects of platelet rich plasma and chondrocyte co-culture on MSC chondrogenesis, hypertrophy and pathological responses. EXCLI Journal. 2017;**16**:1031-1045. DOI: 10.17179/excli2017-453

[4] Xu J, Zhang C. In vitro isolation and cultivation of human chondrocytes for osteoarthritis renovation. In Vitro Cellular & Developmental Biology. Animal. 2014;**50**:623-629. DOI: 10.1007/s11626-014-9742-5

[5] Akuri MC, Barbalho SM, Val RM, Guiguer EL. Reflections about osteoarthritis and *Curcuma longa*. Pharmacognosy Reviews. 2017;**11**:8-12. DOI: 10.4103/phrev.phrev_54_16

[6] Schadow S, Simons VS, Lochnit G, Kordelle J, Gazova Z, Siebert HC, Steinmeyer J. Metabolic response of human osteoarthritic cartilage to biochemically characterized collagen hydrolysates. International Journal of Molecular Sciences. 2017;**18**(1):207. DOI: 10.3390/ijms18010207 pii: E207

[7] Xie XW, Wan RZ, Liu ZP. Recent research advances in selective matrix metalloproteinase-13 inhibitors as anti-osteoarthritis agents. ChemMedChem. 2017;**12**:1157-1168. DOI: 10.1002/cmdc.201700349

[8] Comblain F, Sanchez C, Lesponne I, Balligand M, Serisier S, Henrotin Y. Curcuminoids extract, hydrolyzed collagen and green tea extract synergically inhibit inflammatory and catabolic mediator's synthesis by normal bovine and osteoarthritic human chondrocytes in monolayer. PLoS One. 2015;**10**:e0121654. DOI: 10.1371/journal.pone.0121654

[9] Zhang Z, Leong DJ, Xu L, He Z, Wang A, Navati M, Kim SJ, Hirsh DM, Hardin JA, Cobelli NJ, Friedman JM, Sun HB. Curcumin slows osteoarthritis progression and relieves osteoarthritis-associated pain symptoms in a post-traumatic osteoarthritis mouse model. Arthritis Research & Therapy. 2016;**18**:128. DOI: 10.1186/s13075-016-1025-y

[10] Rannou F, Pelletier JP, Martel-Pelletier J. Efficacy and safety of topical NSAIDs in the management of osteoarthritis: Evidence from real-life setting trials and surveys. Seminars in Arthritis and Rheumatism. 2016;**45**:S18-S21. DOI: 10.1016/j.semarthrit

[11] Li P, Zheng Y, Chen X. Drugs for autoimmune inflammatory diseases: From small molecule compounds to anti-TNF biologics. Frontiers in Pharmacology. 2017;**8**:460. DOI: 10.3389/fphar.2017.00460

[12] Euppayo T, Punyapornwithaya V, Chomdej S, Ongchai S, Nganvongpanit K. Effects of hyaluronic acid combined with anti-inflammatory drugs compared with hyaluronic acid alone, in clinical trials and experiments in osteoarthritis: A systematic review and meta-analysis. BMC Musculoskeletal Disorders. 2017;**18**:387. DOI: 10.1186/s12891-017-1743-6

[13] Del Grossi Moura M, Lopes LC, Biavatti MW, Kennedy SA, de Oliveira E, Silva MC, Silva MT, de Cássia Bergamaschi C. Oral herbal medicines marketed in Brazil for the treatment of osteoarthritis: A systematic review and meta-analysis. Phytotherapy Research. 2017; **31**(11):1676-1685. DOI: 10.1002/ptr.5910

[14] Johnson JL. Metalloproteinases in atherosclerosis. European Journal of Pharmacology. 2017;**816**:93-106. DOI: 10.1016/j.ejphar.2017.09.007 pii: S0014-2999(17)30591-5

[15] Henrotin Y, Clutterbuck AL, Allaway D, Lodwig EM, Harris P, Mathy-Hartert M, Shakibaei M, Mobasheri A. Biological actions of curcumin on articular chondrocytes. Osteoarthritis and Cartilage. 2010;**18**:141-149. DOI: 10.1016/j.joca.2009.10.002

[16] Wojdasiewicz P, Poniatowski LA, Szukiewicz D. The role of inflammatory and anti-inflammatory cytokines in the pathogenesis of osteoarthritis. Mediators of Inflammation. 2014;**2014**:561459. DOI: 10.1155/2014/561459

[17] Yang CY, Chanalaris A, Troeberg L. ADAMTS and ADAM metalloproteinases in osteoarthritis—Looking beyond the 'usual suspects. Osteoarthritis and Cartilage. 2017;**25**:1000-1009. DOI: 10.1016/j.joca.2017.02.791

[18] Chin KY. The spice for joint inflammation: Anti-inflammatory role of curcumin in treating osteoarthritis. Drug Design, Development and Therapy. 2016;**10**:3029-3042. DOI: 10.2147/DDDT.S117432

[19] Yeh CC, YH S, Lin YJ, Chen PJ, Shi CS, Chen CN, et al. Evaluation of the protective effects of curcuminoid (curcumin and bisdemethoxycurcumin)-loaded liposomes against bone turnover in a cell-based model of osteoarthritis. Drug Design, Development and Therapy. 2015;**9**:2285-2300. DOI: 10.2147/DDDT.S78277

[20] Tamma R, Zallone A. Osteoblast and osteoclast crosstalks: From OAF to Ephrin. Inflammation & Allergy Drug Targets. 2012;**11**:196-200

[21] Bharti AC, Takada Y, Aggarwal BB. Curcumin (diferuloylmethane) inhibits receptor activator of NF-kappa B ligand-induced NF-kappa B activation in osteoclast precursors and suppresses osteoclastogenesis. Journal of Immunology. 2004;**172**:5940-5947. DOI: https://doi.org/10.4049/jimmunol.172.10.5940

[22] Wei S, Teitelbaum SL, Wang MW, Ross FP. Receptor activator of nuclear factor-kappa b ligand activates nuclear factor-kappa b in osteoclast precursors. Endocrinology. 2001;**142**:1290-1295. DOI: 10.1210/endo.142.3.8031

[23] Liu YD, Yang HX, Liao LF, Jiao K, Zhang HY, Lu L, et al. Systemic administration of strontium or NBD peptide ameliorates early stage cartilage degradation of mouse mandibular condyles. Osteoarthritis and Cartilage. 2016;**24**:178-187. DOI: 10.1016/j.joca.2015.07.022

[24] Xu L, Guo H, Li C, Xu J, Fang W, Long XA. Time-dependent degeneration manner of condyle in rat CFA-induced inflamed TMJ. American Journal of Translational Research. 2016;**8**:556-567

[25] Mahmood K, Zia KM, Zuber M, Salman M, Anjum MN. Recent developments in curcumin and curcumin based polymeric materials for biomedical applications: A review. International Journal of Biological Macromolecules. 2015;**81**:877-890. DOI: 10.1016/j.ijbiomac.2015.09.026

[26] Yadav SK, Sah AK, Jha RK, Sah P, Shah DK. Turmeric (curcumin) remedies gastroprotective action. Pharmacognosy Reviews. 2013;**7**:42-46. DOI: 10.4103/0973-7847.112843

[27] Aggarwal BB, Harikumar KB. Potential therapeutic effects of curcumin, the anti-inflammatory agent, against neurodegenerative, cardiovascular, pulmonary, metabolic, autoimmune and neoplastic diseases. The International Journal of Biochemistry & Cell Biology. 2009;**41**:40-59. DOI: 10.1016/j.biocel.2008.06.010

[28] A M, Henrotin Y, Biesalski HK, Shakibaei M. Scientific evidence and rationale for the development of curcumin and resveratrol as nutraceuticals for joint health. International Journal of Molecular Sciences. 2012;**13**:4202-4232. DOI: 10.3390/ijms13044202

[29] Prasad S, Tyagi AK. Curcumin and its analogues: A potential natural compound against HIV infection and AIDS. Food & Function. 2015;**6**:3412-3419. DOI: 10.1039/c5fo00485c

[30] J Wang, J Ma, JH Gu, FY Wang, XS Shang, HR Tao, X Wang. Regulation of type II collagen, matrix metalloproteinase-13 and cell proliferation by interleukin-1β is mediated by curcumin via inhibition of NF-κB signaling in rat chondrocytes. Molecular Medicine Reports. 2017;**16**:1837-1845. DOI: 10.3892/mmr.2017.6771

[31] Oliviero F, Scanu A, Zamudio-Cuevas Y, Punzi L, Spinella P. Anti-inflammatory effects of polyphenols in arthritis. Journal of the Science of Food and Agriculture. 2017. DOI: 10.1002/jsfa.8664

[32] Agrawal DK, Mishra PK. Curcumin and its analogues: Potential anticancer agents. Medicinal Research Reviews. 2010;**30**:818-860. DOI: 10.1002/med.20188

[33] Wang XP, Wang QX, Lin HP, Chang N. Anti-tumor bioactivities of curcumin on mice loaded with gastric carcinoma. Food & Function. 2017;**8**(9):3319-3326. DOI: 10.1039/c7fo00555e

[34] Feng T, Wei Y, Lee RJ, Zhao L. Liposomal curcumin and its application in cancer. International Journal of Nanomedicine. 2017;**21**(12):6027-6044. DOI: 10.2147/IJN.S132434

[35] Xu X, Zhu Y. Curcumin inhibits human non-small cell lung cancer xenografts by targeting STAT3 pathway. American Journal of Translational Research. 2017;**9**(8):3633-3641

[36] Saberi-Karimian M, Katsiki N, Caraglia M, Boccellino M, Majeed M, Sahebkar A. Vascular endothelial growth factor: An important molecular target of curcumin. Critical Reviews in Food Science and Nutrition. 2017;**30**:1-14. DOI: 10.1080/10408398.2017.1366892

[37] Riva A, Togni S, Giacomelli L, Franceschi F, Eggenhoffner R, Feragalli B, Belcaro G, Cacchio M, Shu H, Dugall M. Effects of a curcumin-based supplementation in asymptomatic subjects with low bone density: A preliminary 24-week supplement study. European Review for Medical and Pharmacological Sciences. 2017;**21**:1684-1689

[38] Chen Z, Xue J, Shen T, Mu S, Fu Q. Curcumin alleviates glucocorticoid-induced osteoporosis through the regulation of the Wnt signaling pathway. International Journal of Molecular Medicine. 2016;**37**:329-338. DOI: 10.3892/ijmm.2015.2432

[39] Flores G. *Curcuma longa* L. extract improves the cortical neural connectivity during the aging process. Neural Regeneration Research. 2017 Jun;**12**:875-880. DOI: 10.4103/1673-5374.208542

[40] Mittelstaedt D, Xia Y. Depth-dependent glycosaminoglycan concentration in articular cartilage by quantitative contrast-enhanced micro–computed tomography. Cartilage. 2015;**6**:216-225. DOI: 10.1177/1947603515596418

[41] Lovell S, Burchell RK, Roady PJ, Fredrickson RL. Gal. Canine intrathoracic sarcoma with ultrastructural characteristics of human synovial sarcoma - case report. BMC Veterinary Research. 2017;**13**:247. DOI: 10.1186/s12917-017-1181-6

[42] Hao L, Wan Y, Xiao J, Tang Q, Deng H, Chen L. A study of Sirt1 regulation and the effect of resveratrol on synoviocyte invasion and associated joint destruction in rheumatoid arthritis. Molecular Medicine Reports. 2017;**16**:5099-5106. DOI: 10.3892/mmr.2017.7299

[43] Benedetti G, Bonaventura P, Lavocat F, Miossec P. IL-17A and TNF-α increase the expression of the antiapoptotic adhesion molecule amigo-2 in arthritis synoviocytes. Frontiers in Immunology. 2016;**7**:254. DOI: 10.3389/fimmu.2016.00254

[44] Buckwalter JA, Mankin HJ. Articular cartilage: Degeneration and osteoarthritis, repair, regeneration, and transplantation. Instructional Course Lectures. 1998;**47**:487-504

[45] Mobasheri A, Airley R, Foster CS, Schulze-Tanzil G, Shakibaei M. Post-genomic applications of tissue microarrays: Basic research, prognostic oncology, clinical genomics and drug discovery. Histology and Histopathology. 2004;**19**:325-335. DOI: 10.14670/HH-19.325.

[46] Mollazadeh H, Cicero AFG, Blesso CN, Pirro M, Majeed M, Sahebkar A. Immune modulation by curcumin: The role of interleukin-10. Critical Reviews in Food Science and Nutrition. 2017;**11**:1-13. DOI: 10.1080/10408398.2017.1358139

[47] Jagetia GC, Aggarwal BB. "spicing up" of the immune system by curcumin. Journal of Clinical Immunology. 2007;**27**:19-35. DOI: 10.1007/s10875-006-9066-7

[48] Buhrmann C, Mobasheri A, Matis U, Shakibaei M. Curcumin mediated suppression of nuclear factor-kB promotes chondrogenic differentiation of mesenchymal stem cells in a high-density co-culture microenvironment. Arthritis Research & Therapy. 2010;**12**:R127. DOI: 10.1186/ar3065

[49] Saklatvala J. Tumour necrosis factor alpha stimulates resorption and inhibits synthesis of proteoglycan in cartilage. Nature. 1986;**322**:547-549. DOI: 10.1038/322547a0

[50] Séguin CA, Bernier SM. TNFalpha suppresses link protein and type II collagen expression in chondrocytes: Role of MEK1/2 and NF-kappaB signaling pathways. Journal of Cellular Physiology. 2003;**197**:356-369. DOI: 10.1002/jcp.10371

[51] Perkins K, Sahy W, Beckett RD. Efficacy of *Curcuma* for treatment of osteoarthritis. Journal of Evidence-Based Complementary & Alternative Medicine. 2017;**22**:156-165. DOI: 10.1177/2156587216636747

[52] Panahi Y, Alishiri GH, Parvin S, Sahebkar A. Mitigation of systemic oxidative stress by curcuminoids in osteoarthritis: Results of a randomized controlled trial. Journal of Dietary Supplements. 2016;**13**:209-220. DOI: 10.3109/19390211.2015.1008611

[53] Kuptniratsaikul V, Dajpratham P, Taechaarpornkul W, Buntragulpoontawee M, Lukkanapichonchut P, Chootip C, et al. Efficacy and safety of *Curcuma domestica* extracts compared with ibuprofen in patients with knee osteoarthritis: A multicenter study. Clinical Interventions in Aging. 2014;**9**:451-458. DOI: 10.2147/CIA.S58535

[54] Appelboom T, Maes N, Albert A. A new *Curcuma* extract (flexofytol®) in osteoarthritis: Results from a Belgian real-life experience. Open Rheumatology Journal. 2014;**8**:77-81. DOI: 10.2174/1874312901408010077

[55] Kizhakkedath R. Clinical evaluation of a formulation containing *Curcuma longa* and *Boswellia serrata* extracts in the management of knee osteoarthritis. Molecular Medicine Reports. 2013;**8**:1542-1548. DOI: 10.3892/mmr.2013.1661

[56] Madhu K, Chanda K, Saji MJ. Safety and efficacy of *Curcuma longa* extract in the treatment of painful knee osteoarthritis: A randomized placebo-controlled trial. Inflammopharmacology. 2013;**21**:129-136. DOI: 10.1007/s10787-012-0163-3

[57] Kertia N, Asdie AH, Rochmah W, Marsetyawan. Ability of curcuminoid compared to diclofenac sodium in reducing the secretion of cycloxygenase-2 enzyme by synovial fluid's monocytes of patients with osteoarthritis. Acta Medica Indonesiana. 2012;**44**:105-113

[58] Chopra A, Lavin P, Patwardhan B, Chitre D. A 32-week randomized, placebo-controlled clinical evaluation of RA-11, an Ayurvedic drug, on osteoarthritis of the knees. Journal of Clinical Rheumatology. 2004;**10**:236-245. DOI: 10.1097/01.rhu.0000138087.47382.6d

[59] Chandrasekaran CV, Sundarajan K, Edwin JR, Gururaja GM, Mundkinajeddu D, Agarwal A. Immune-stimulatory and anti-inflammatory activities of *Curcuma longa* extract and its polysaccharide fraction. Pharmacognosy Research. 2013;**5**:71-79. DOI: 10.4103/0974-8490.110527

[60] Szymusiak M, Hu X, Leon Plata PA, Ciupinski P, Wang ZJ, Liu Y. Bioavailability of curcumin and curcumin glucuronide in the central nervous system of mice after oral delivery of nano-curcumin. International Journal of Pharmaceutics. 2016;**511**:415-423. DOI: 10.1016/j.ijpharm.2016.07.027

[61] Zamarioli CM, Martins RM, Carvalho EC, Freitas Luis AP. Nanoparticles containing curcuminoids (*Curcuma longa*): Development of topical delivery formulation. Revista Brasileira de Farmacognosia. 2015;**25**:53-60. DOI: 10.1016/j.bjp.2014.11.010

[62] Mirzaei H, Naseri G, Rezaee R, Mohammadi M, Banikazemi Z, Mirzaei HR, et al. Curcumin: A new candidate for melanoma therapy. International Journal of Cancer. 2016;**139**:1683-1695. DOI: 10.1002/ijc.30224

[63] Gupta NK, Dixit VK. Bioavailability enhancement of curcumin by complexation with phosphatidyl choline. Journal of Pharmaceutical Sciences. 2011;**100**:1987-1995. DOI: 10.1002/jps.22393

[64] Bhardwaj RK, Glaeser H, Becquemont L, Klotz U, Gupta SK, Fromm MF. Piperine, a major constituent of black pepper, inhibits human P-glycoprotein and CYP3A4. The Journal of Pharmacology and Experimental Therapeutics. 2002;**302**:645-650. DOI: 10.1124/jpet.102.034728

Chapter 4

Cell Therapy and Tissue Engineering for Cartilage Repair

María Piñeiro-Ramil, Rocío Castro-Viñuelas,
Clara Sanjurjo-Rodríguez, Tamara Hermida-Gómez,
Isaac Fuentes-Boquete, Francisco J. de Toro-Santos,
Francisco J. Blanco-García and Silvia M. Díaz-Prado

Additional information is available at the end of the chapter

http://dx.doi.org/10.5772/intechopen.70406

Abstract

The integrity of the articular cartilage is necessary for the proper functioning of the diarthrodial joint. The self-repair capacity of this tissue is very limited and, currently, there is no effective treatment capable of restoring it. The degradation of the articular cartilage leads to osteoarthritis (OA), a leading cause of pain and disability mainly among older people.

Different cell treatments have been developed with the aim of forming a repair tissue with the characteristics of native articular cartilage, including cellular therapy and tissue engineering. Cell therapy-based approaches include bone marrow-stimulating techniques, implants of periosteum and perichondrium, ostechondral grafting and implantation of chondrogenic cells as chondrocytes, mesenchymal stem cells or induced pluripotent stem cells. In tissue engineering-based approaches cell-free scaffolds capable of recruiting endogenous cells or chondrogenic cell-loaded scaffolds may be used.

However, despite the numerous treatments available nowadays, no technique has been able to consistently regenerate native articular cartilage in clinical trials. Although many cell therapy and tissue engineering studies have shown promising results and clinical improvement, these treatments generate a fibrocartilaginous tissue different from native articular cartilage. More research is needed to improve cell-based approaches and prove its efficacy

Keywords: regenerative medicine, chondrogenic cells, mesenchymal stem cells (MSCs), induced pluripotent stem cells (iPS), scaffolds

1. Introduction

The integrity of the structure of the articular cartilage is necessary for the proper functioning of the diarthrodial joint. Articular hyaline cartilage provides a resistant, smooth, and lubricated surface, which avoids friction between bones. Thus, hyaline cartilage absorbs and minimizes the pressures produced in the movement of the joint, allows bones to glide over one another with minimal friction, and facilitates the coupling between articular surfaces. Due to its elasticity, articular cartilage absorbs an important part of the compression force, reducing the load supported by the underlying bone structure [1–3].

Traditionally, osteoarthritis (OA) was defined as a degenerative joint disease, characterized by the alteration in the integrity of the articular cartilage [1]. Nowadays, it is known that although the degradation of articular cartilage is the central event in the pathogenesis of OA, synovial tissue and subchondral bone also participate in the onset and development of this disease [4]. The degree of compromise of these components of the joint leads not only to variability between the clinical profiles of patients, but also between different joints of the same patient [5]. On this basis, the Osteoarthritis Research Society International (OARSI) has defined OA as a heterogeneous disorder of movable joints, manifested as genetic, metabolic, and inflammatory changes in the joint, as well as anatomic and/or physiological conditions that may lead to the symptoms associated with the disease. OA is characterized by cell stress and extracellular matrix degradation initiated by micro- and macro-injury that activates maladaptive repair responses including pro-inflammatory pathways of innate immunity [6]. OA is one of the most common chronic health conditions and a leading cause of pain and disability among adults [2, 7]. OA is one of the most prevalent diseases in older people and its incidence, which increases with age, is expected to rise along with the median age of the population [3, 8].

The self-repair capacity of articular cartilage is very limited as it is an avascular and aneural tissue. Due to this absence of vascularity, progenitor cells present in blood and marrow cannot enter into the damaged region to influence or contribute to the reparative process [9, 10]. In addition, because of aneurality, chondral lesions are not detected, and thus patients are not medically treated until more severe lesions are formed [11, 12].

Currently, there is no effective treatment capable of restoring the physiological properties of the osteochondral unit (**Figure 1A**) [13, 14] and the prosthetic replacement is necessary at the final clinical stage (**Figure 1B**) [6]. Different cell treatments have been developed with the aim of forming a repair tissue with structural, biochemical, and functional characteristics equivalent to those of native articular cartilage (**Figure 2**). Scientists have sought several different ways to repair articular cartilage after traumatic damage, which can lead to secondary OA or degeneration of the cartilage [13, 15–17].

It is necessary to highlight that "repair" refers to the restoration of a damaged articular surface with the formation of a neocartilage tissue, which resembles to the native cartilage and "regeneration" refers to the formation of a tissue indistinguishable from the native articular cartilage [16]. Cellular therapy (using cells) and tissue engineering (combining cells, scaffolds, and bioactive factors) have emerged as alternative clinical approaches. However, despite the numerous treatments available nowadays, no technique has been able to consistently regenerate normal hyaline cartilage in clinical trials [3, 18]. Long-term follow-up studies are expected to be performed in the coming years to confirm safety and effectiveness of these new approaches [3].

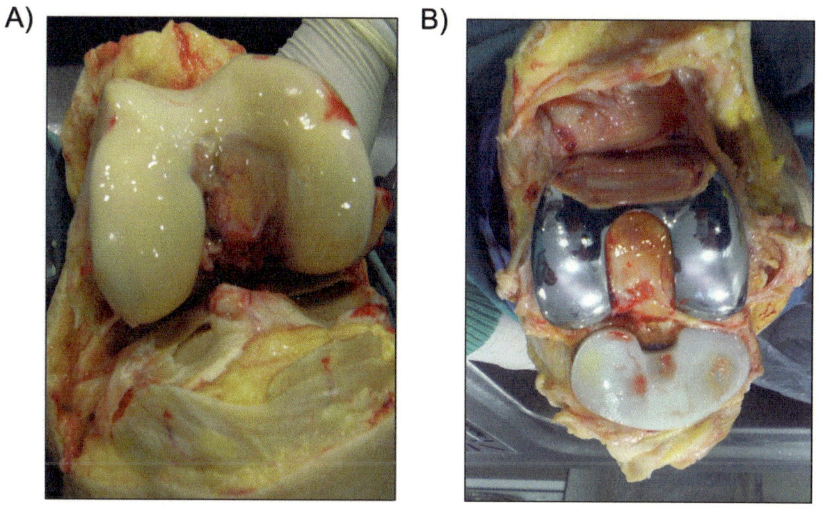

Figure 1. Images showing (A) healthy knee joint and (B) prosthetic joint replacement.

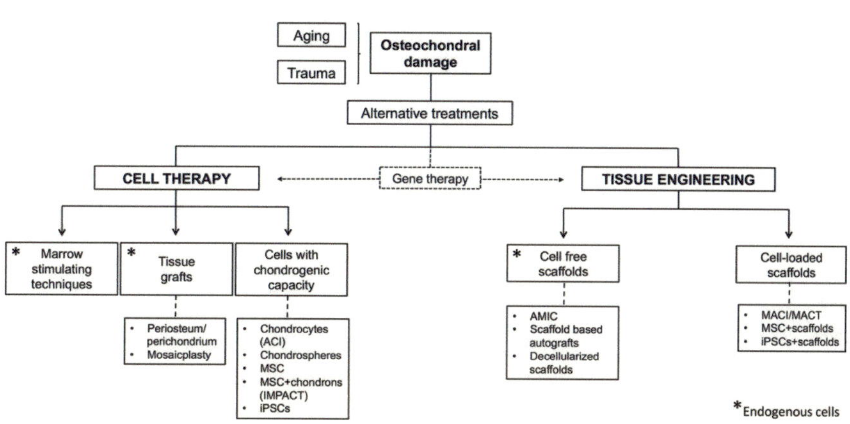

Figure 2. Diagram showing an overview of the alternative treatments for osteochondral damage.

2. Cell therapy

Cell therapy is a relatively new approach based on the regeneration or repair of a damaged tissue using autologous or allogenic cells.

2.1. Marrow stimulating techniques

Bone marrow stimulating techniques (MSTs) are based on the use of endogenous mesenchymal stromal cells (MSCs). This type of technique is used in the treatment of chondral lesions with less of 15 mm of diameter [19].

Penetration of subchondral bone is among the oldest and still the most commonly used method to stimulate regeneration of neocartilage [16, 20]. Arthroscopic techniques like drilling, abrasion arthroplasty or microfracture are different tools to perforate the subchondral bone [12], allowing MSCs and growth factors from the bone marrow to infiltrate the lesion [15]. A blood clot is formed in the defect, acting as a scaffold and mediating the inflammatory response (through cytokines) [19].

However, it was described that endogen bone marrow angiogenic factors favor osteogenesis, instead of chondrogenesis, of bone marrow MSCs [11]. Generated repair tissue frequently ends up degenerating [21] and usually presents type I collagen (fibrocartilage phenotype) and lacks hyaline cartilage viscoelastic properties [22].

2.2. Tissue grafts

Tissue grafts have potential benefits in cartilage repair since they contain cell populations with chondrogenic capacity.

2.2.1. Implants of periosteum and perichondrium

In the 90s, autologous strips of perichondrium were used to treat chondral defects [23, 24]. Periosteum and perichondrium contain MSCs that are capable of chondrogenesis and act as a biological membrane [16]. However, the ability of periosteum MSCs to proliferate and differentiate into chondrocytes decreases with age [25].

The clinical outcomes of perichondrium implants are similar to those of subchondral perforation [26]. Calcification of the periosteum grafts had been mentioned as a problem in the long term [16].

2.2.2. Mosaicplasty

Autologous mosaicplasty is widely used for treating chondral and osteochondral defects. The most used technique is the osteochondral autologous transplantation (OAT), which consists in the translocation of osteochondral cylinders from not loading areas to the affected areas of the joint [15].

Even though good to excellent short-term subjective results were obtained, clinical and radiological midterms to long-term outcomes of mosaicplasty were moderate. Further limitations are donor-site morbidity, technical difficulty, special equipment, lesion size, and fibrocartilaginous repair [16, 27]. OAT might be more appropriate for lesions smaller than 2–3 cm^2 [28].

Another problem is the lack of congruence between the osteochondral cylinders implanted and the lesion area, and the differences in cartilage height of the defect and surrounding native cartilage, altering the distribution of stress and compression forces [16, 27].

Allogenic mosaicplasty has shown successful outcomes and its main advantage over autograft transplantation is the lack of donor-site morbidity. Nevertheless, the amount of transplanted bone has to be minimum because the allograft failure is mostly due to collapse of the subchondral bone [22].

Nowadays, synthetic cylindrical plugs for implant similar to OAT exist but studies have shown universal failure to incorporate these plugs into the subchondral bone, with formation of cysts [22].

In addition to fresh osteochondral grafts, particulated cartilage grafts, which are formed by combining fragments of cartilage with fibrin glue, may also be used. Superficial chondrocytes, released from the extracellular matrix as a consequence of the fragmentation of the cartilage, produce additional extracellular matrix that integrates the particulate graft with native cartilage and fills the defect [29].

2.3. Implantation of cells with chondrogenic capacity

Chondrogenic potential of different cell types (**Figure 3**) was tested for hyaline cartilage repair.

2.3.1. Autologous chondrocyte implantation

The autologous chondrocyte implantation (ACI) was firstly described by Peterson et al. [30]. This technique consists of harvesting a cartilage piece from a low-weight-bearing area of the joint and culture-expanding the chondrocytes to implant into the lesion. The lesion is sealed with autologous periosteum to avoid cell loss.

ACI is only applicable to small size (3–4 cm^2) focal lesions surrounded by healthy cartilage [15, 28]. Other limitations are dedifferentiation of chondrocytes during culture expansion, the low amount of chondrocytes obtained and multiple surgical procedures involved [31, 32]. Further, donor-site morbidity of cartilage and bone for chondrocyte and periosteum obtaining was observed [15, 33, 34].

ACI is considered superior to MSTs regarding the quality of the repaired tissue, although there are conflicting results [28].

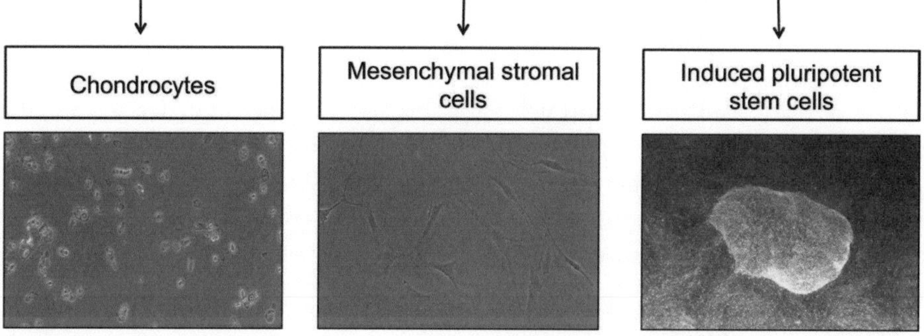

Figure 3. Diagram showing the different cell sources, most commonly used in cartilage treatment using cell therapy: chondrocytes (left), mesenchymal stromal cells (middle), and induced pluripotent stem cells (right).

2.3.2. *Chondrospheres*

The technique of chondrospheres consists of the generation and implantation of spheroids of autologous or allogenic articular chondrocytes [29]. Autologous chondrocytes are obtained from undamaged articular cartilage, expanded *in vitro,* and condensed in order to form spheroids, which then are coalesced. Chondrospheres have shown to be able to adhere, integrate into hyaline cartilage defect and produce cartilaginous extracellular matrix in mouse, mini pig, and horse cartilage defect models, as well as in artificial defects in human cartilage explants [35–37]. A phase III clinical trial is currently ongoing in Germany and Poland to investigate the efficacy of this technology compared to microfracture in the treatment of cartilage defects of knee joints [38].

2.3.3. *Mesenchymal stromal cells*

Human MSCs are nonhematopoietic multipotent progenitor cells with long-term self-renewal ability and the capacity to differentiate along multiple cell lineages, including cartilage, as well as immunomodulatory features [39–41]. MSCs are responsible for normal tissue renewal and for response to injury and may be an alternative to chondrocytes for the development of new therapeutic approaches for the treatment of cartilage defects.

In vitro and *in vivo* studies of clonally derived MSCs demonstrated that these cells consist of subsets that present different surface markers expression and different capacities for cellular differentiation [42]. These cells are considered a potential cell source for cell therapy since they can be easily collected from various tissues such as bone marrow [43], adipose tissue [44], synovial membrane [42], and amniotic membrane [45], among others. However, the equivalence of chondrogenic differentiation potential of MSCs derived from different tissues is a matter of considerable debate [46].

For cell therapy approaches, either autologous or allogenic MSCs can be used. MSCs do not express major histocompatibility complex class II (MHC II) and its co-stimulatory molecules, and barely express major histocompatibility complex class I (MHC I), so that they do not produce alloreactivity, avoiding rejection problems. This feature turns MSCs into a feasible cell source for allogenic transplantation [40, 47].

The therapeutic potential of autologous MSCs derived from different tissues to stimulate the regeneration of cartilage in OA has been reported in several preclinical studies [48, 49]. Bone marrow-derived MSCs suspended in hyaluronic acid and administrated by intra-articular injection have been used to promote cartilage repair in animal models such as guinea pig, mini pig, goat and donkey, leading to improvement in cartilage regeneration, less cartilage destruction and reduced osteophyte formation [50–53]. MSCs derived from other sources have also been used; for example, transplantation of synovial MSCs was used to repair osteochondral defects in rabbits [54], and intra-articular injection of adipose-derived MSCs was used to treat chronic osteoarthritis in dogs, showing significant improvement in MSCs-treated joints [55].

One of the MSCs transplantation techniques for cartilage focal lesions is a variation of ACI in which bone marrow MSCs are injected into defects and closed with periosteal membrane to be differentiated toward chondrocytes [56]. The first clinical study using MSCs to treat OA was performed by Wakitani et al. [57]. In this study, bone marrow-derived MSCs were transplanted into the articular cartilage defect and covered with autologous periosteum. Although the arthroscopic and histological grading score was better in the cell-transplanted group than in the control one, the clinical improvement was not very clear. Since then, several clinical studies have been performed, mainly using intra-articular injection of autologous bone marrow-derived MSCs, showing some degree of improvement in terms of clinical outcomes and repaired cartilage tissue quality [58–60]. However, several studies described a lack of engraftment into cartilage defects [61] and it is important to highlight that most of the clinical trials are I and I/II phases, indicating the immaturity of MSC clinical applications in OA [49].

Limitations of this approach are that culture expansion is not avoided, cell yield is often low and MSCs differentiation capacity decreases with age of the donor [21]. This is a problem in regenerative therapies for degenerative diseases such as OA, where most of patients are aged [61]. Given that the age of patient and the size of the lesion affect the outcome, the cut-off points for the risk of failure have been suggested at age greater than 60 years and lesion size larger than 6.0 cm^2 [28].

2.3.4. Mesenchymal stromal cells combined with autologous chondrons

A novel cell therapy approach is based on combining autologous chondrocytes in their pericellular matrix (chondrons) and allogenic MSCs, which was called Instant MSC Product accompanying Autologous Chondron Transplantation (IMPACT) and performed by De Windt et al. [62]. In this phase I clinical trial, patients with focal cartilage defects were treated using a mix of 80–90% allogenic MSCs and 10–20% autologous chondrons combined with fibrin glue. In this approach, chondrons are "recycled" from debrided cartilage instead of being harvested from a low-weight-bearing area of the joint, as occurring in ACI. The combination of this recycled chondrons with allogenic human bone marrow MSCs stimulates cartilage regeneration and provides clinical improvement. Surprisingly, although the co-implantation of chondrons and MSCs provides better results in comparison with implantation of chondrons or MSCs alone [63], no allogenic cells were detected in the repaired cartilage after 1 year, suggesting that MSCs have trophic effects that stimulate chondrons to regenerate cartilage. The quality of the repaired tissue and the clinical outcome using the IMPACT technique was similar or even superior in comparison with ACI. Furthermore, IMPACT technique presents the advantage of allowing to perform both surgeries on the same day (the extraction of cartilage and the implantation of cells) [62].

2.3.5. Induced pluripotent stem cells

Pluripotent cells could provide an unlimited and renewable cell source that can be induced to differentiate into any cell type. In fact, pluripotent cells of embryonic origin [61, 64], embryonic human stem cells (hESCs), or induced to pluripotency [65], induced pluripotent stem cells (iPSCs), have shown to produce cartilage under specific conditions. iPSCs have been generated from adult cells (**Figure 4A**) using defined factors (**Figure 4B**) [66]. These cells present similar morphology (**Figure 4C**), proliferation capacity, genetic expression and epigenetic pattern, and pluripotency characteristics to hESCs [66, 67].

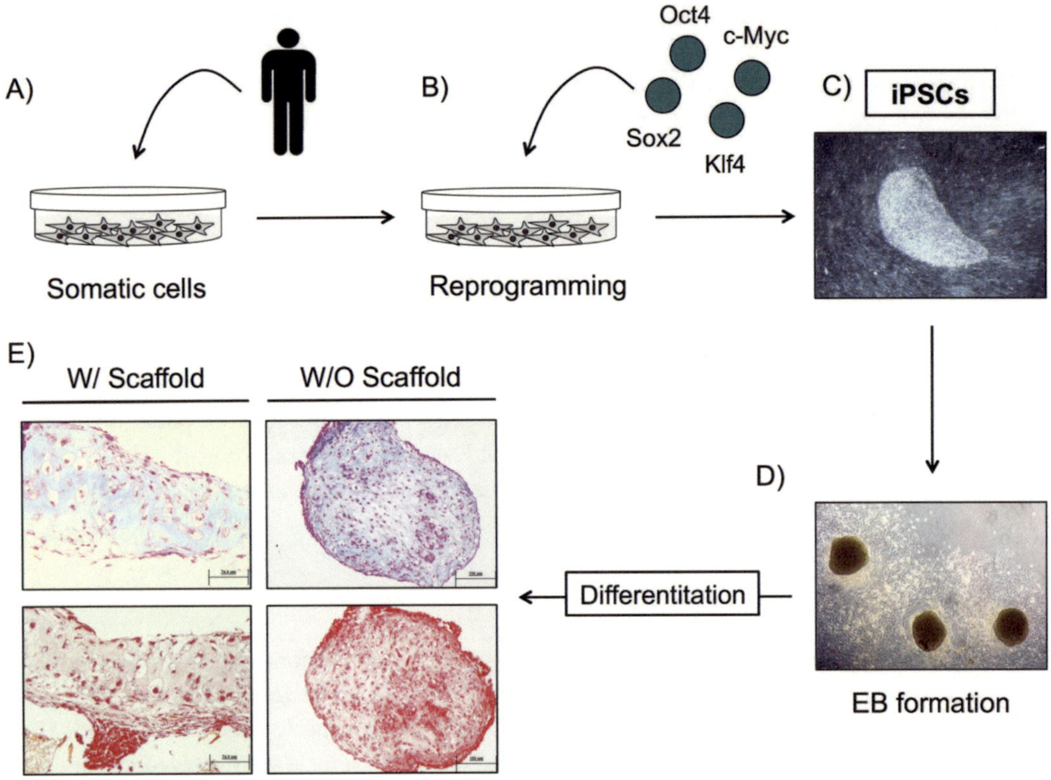

Figure 4. Scheme representing the role of iPSCs in tissue engineering. (A) Harvesting somatic cells from the patient. (B) Reprogramming the cells using the factors Oct4, Sox2, Klf4, and c-Myc. (C) iPSc colony obtained after reprogramming. (D) Embryoid bodies (EB) formation. (E) Differentiation of the iPSc toward chondrocytes with (W/) or without (W/O) scaffold.

iPSCs seem to be an alternative tool to chondrocytes for cartilage repair as they can be expanded before starting their differentiation (using or not embryoid bodies formation) toward chondrocytes (**Figure 4D**). Then, iPSC-derived chondrocytes can be cultured in three-dimensional culture with scaffold (**Figure 4E**, w/Scaffold), or cultured without a scaffold (**Figure 4E**, w/o Scaffold), to create cartilaginous tissues *in vitro* before transplantation to repair large defects [68].

In addition, iPSCs seem to be an alternative tool to MSCs for cartilage repair. After *in vitro* chondrogenesis, iPSCs showed lower hypertrophic markers than MSCs [69]. The risk of iPSCs teratoma formation in cell therapy or tissue engineering can be avoided using pre-differentiated cells before implantation [70, 71]. Also, the use of iPSCs avoids the problem of *in vivo*-age-dependent and *in vitro*-passage-dependent MSC senescence [72].

Yamashita et al. [73] optimized a protocol of chondrogenic differentiation using human iPSCs to form homogenous cartilaginous particles. After the transplantation of these chondrogenic particles into joint surface defects in immunodeficient rats and immunosuppressed mini pigs, they observed cartilaginous neotissue with potential for integration into native cartilage.

Nowadays, there are no clinical studies published about cartilage cell therapy using iPSCs. Although cell therapy or tissue engineering using iPSCs are promising tools, their clinical use is not legalized either by the scientific community or by existing international legislation yet, except in Japan.

3. Tissue engineering

The lack of efficient treatments for cartilage repair motivates the researchers to develop, by tissue engineering, biological tissue substitutes that can be implanted to replace the affected area of the joint [74]. Tissue engineering is not widespread yet in surgical procedures, although there are many combinations of different cells and supports being tested both *in vitro* and *in vivo*.

In this way, different strategies were developed for cartilage regeneration, based on the use of scaffolds and endogenous or exogenous cells. Whereas in *in vitro* studies scaffolds are usually combined with cells and bioactive factors, in most *in vivo* studies the scaffolds are used only combined with cells because those factors are present in the joint (e.g., AMIC described below).

In vitro administration of growth factors (transforming growth factor 1 or 3, bone morphogenetic proteins 2 or 7, and insulin growth factor 1, among others) have been used to induce chondrogenic differentiation of MSCs and iPSCs. However, the effect of application of these molecules is dose, timing of administration and cell type-dependent [75]. That is why, in recent years, scaffolds were functionalized with bioactive factors or other molecules for *in vivo* cartilage therapies, as a delivery system [76] or stimulation for MSCs. For example, the addition of proteoglycans to collagen biomaterials had improved bone marrow MSCs chondrogenic differentiation [43, 77].

A broad variety of biomaterials have been successfully developed to support proliferation, infiltration, or differentiation of allogeneic transplanted or endogenous MSCs to achieve functional tissue restoration [78]. Scaffolds/biomaterials should be a porous three-dimensional matrix that allow cell migration, adhesion and growth, and support the organization of the growing tissue [79].

However, despite the diffusion of new tissue-engineering techniques and the high number of scaffolds that have been developed and investigated for cartilage regeneration, the ideal matrix material has not been identified yet. Cartilage-engineering strategies have produced promising *in vitro* data, seeding chondrogenic cells on biomaterials with growth factors. However, thus far, no approach has led to the generation of long-term *in vivo* replacement tissue identical to native hyaline cartilage. There are different factors for the lack of stable functional tissue as inflammatory stress or biophysical stimuli [80].

3.1. Cell-free scaffolds and endogenous cells

Cell-free scaffolds are developed for one stage procedure techniques, since they can be implanted alone to attract the endogenous cells. In this case, the aim of using scaffolds is to obtain a suitable microenvironment to recruit and mobilize the host cells, from either the blood or a tissue specific (bone marrow, synovial fluid…) niche for self-repair. Several studies have detected the recruitment of endogenous synovial cells [81, 82] or exogenous-injected MSCs [50] in injured areas after the implantation of empty scaffolds.

Implantation of cell-free scaffolds avoids the issues around the *in vitro* cell culture process, as exogenous cell transplantation is not required. However, clinical results after implantation of cell-free scaffolds for OA treatment are few [3].

3.1.1. Autologous matrix-induced chondrogenesis

The autologous matrix induced chondrogenesis (AMIC) is a second generation MSTs. This is a one-step procedure combining subchondral microfracture with the attachment of a collagen scaffold to the lesion. The initially formed blood clot as produced by microfracturing is protected by the collagen scaffold [83]. The collagen scaffold is thought to stabilize the blood clot, helping to promote early mechanical stability and cartilage regeneration [29]. More complex scaffolds have also been tested in AMIC studies, for example, a biphasic scaffold consisting of calcium triphosphate in the osseous region and poly(lactic-co-glycolic acid) in the cartilaginous region [84].

Even though donor-site morbidity due to removal of periosteum from tibia is avoided, AMIC has similar clinical outcomes to ACI [85].

3.1.2. Scaffold-based autografts

Another approach is the use of scaffold-based autografts, in which harvested cartilage is mechanically minced and uniformly affixed to a biodegradable scaffold, using fibrin glue; then, the scaffold with the cartilage fragments is transferred to the lesion. When compared to microfracture, this scaffold-based autograft procedure resulted in an improvement of functional outcomes and cartilage development [86].

3.1.3. Decellularized extracellular matrix scaffolds

Decellularized extracellular matrix may be used as a scaffold with the potential to retain the bioactive factors needed to support specific tissue formation at the implantation site [87]. Cartilage matrix can be harvested from allogenic sources, then decellularized and used as a scaffold. This approach leads to the improvement of neocartilage formation in preclinical models, in comparison with the living-cartilage implantation [88]. One of the drawbacks of this technique is that the protocols required to decellularization of cartilage also imply some degree of destruction of extracellular matrix components [89]. Decellularized cartilage matrix has been used to treat osteochondral defects in a horse model, obtaining repair of both the bone and cartilage phases [87]. Beside the tissue decellularization, extracellular matrix scaffolds can also be obtained from cultured cells [90].

3.2. Cell-loaded scaffolds

3.2.1. Matrix-associated chondrocyte transplantation

The matrix-associated chondrocyte implantation/transplantation (MACI or MACT) is a second generation ACI, which includes the employment of a bilayer collagen membrane [91]. Essentially, the concept is based on the use of biodegradable polymers as temporary scaffolds for *in vitro* growth of cells and their subsequent transplantation into the defect site. In this case, autologous chondrocytes are previously seeded in the scaffold before implantation into the lesion [12, 83]. Other types of scaffolds (hydrogels, fibrous scaffolds, decellularized ECM, or composites) were later used [85].

MACI presents lower rates of graft hypertrophy than first-generation ACI [92].

3.2.2. Mesenchymal stromal cells on scaffolds

Wakitani et al. [93] observed that MSCs embedded in a collagen gel could differentiate in *in vivo* animal models. Since these first studies, thousands of works were carried out using different types of scaffolds (hydrogels, sponges…), cells, and approaches for chondrogenic scaffolding.

Several *in vivo* studies tried to replicate the distinct osteochondral zones using tri- or bi-layered scaffolds of different composition and/or bioactive factors combined with MSCs. MSCs combined with scaffolds appear to engraft and contribute to cartilage repair, while MSCs injected as a free suspension into the joint do not engraft into the cartilage [61]. This happens because scaffolds can transport cells into the lesion and provide the proper environment for cell differentiation [75, 94].

It was described that cartilage tissue engineering from differentiation-induced *in vitro* MSCs has an inferior quality to that engineered from chondrocytes [95]. However, human amniotic MSCs with human amniotic membrane (as scaffold) showed better reparation in an *in vitro* repair model when compared with bone marrow MSCs and chondrocytes, and demonstrated good adhering capacity to the native cartilage [45]. Also, our group obtained good results using bone marrow MSCs and collagen/heparan sulfate scaffolds in an *in vitro* repair model (**Figure 5**) [96].

3.2.3. Induced pluripotent stem cells on scaffolds

Although tissue-engineering studies using iPSCs are scarce, several studies have shown their potential in chondral repair [21]. Liu et al. [48] have tested the chondrogenesis of murine cells derived from single embryoid bodies. After seeding these cells on polycaprolactone/gelatin scaffolds, they showed a good chondrogenic capacity.

Nowadays, 3D bioprinting into cartilage using iPSCs and bioinks (that act as scaffolds) is being developed [97].

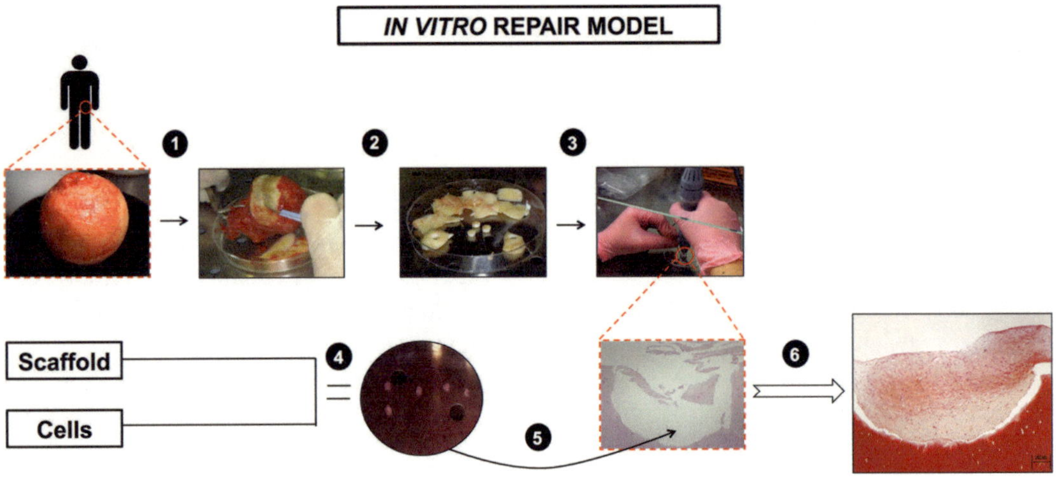

Figure 5. Scheme representing different steps during the development of an *in vitro* cartilage repair model. These steps are on one hand (1) to harvest cartilage explants from the joint (hip), (2) make cartilage punches, and (3) generate the lesion with a driller. On the other hand, (4) to seed the cells on the scaffold and (5) introduce the construct inside the lesion. (6) Safranin O staining showing the final result of the repair model after culture in chondrogenic medium during 2 months.

4. Gene therapy

Gene therapy involves the over-expression of the appropriate gene (anabolic factors, chondroinductor, or anti-inflammatory molecules) and cell type (chondrocytes or chondrogenic cells) for their use in cell therapy and tissue engineering.

Nowadays, no gene products have been approved for OA treatment and few clinical trials have been conducted. At present, only TGF-β gene therapy has been clinically investigated in USA and Korea [3].

5. Conclusions

Although many studies of cell therapy and tissue engineering have shown clinical and functional improvement in joints, these treatments generate a fibrocartilaginous tissue that is different from hyaline articular cartilage. The ability to regenerate articular cartilage that resists the degeneration process still remains elusive.

Acknowledgements

The authors would like to acknowledge CIBER-BBN; Rede Galega de Terapia Celular, Xunta de Galicia (R2014/050); Grupos con Potencial de Cremento, Xunta de Galicia (RTC-2016-5386-1); Unión Europea y Fondo Social Europeo; MINECO-FEDER (RTC-2016-5386-1); Fundación Española de Reumatología (2014 grant); Universidade da Coruña; Fundación Profesor Novoa Santos; Deputación da Coruña; Opocrin S.P.A. (Bruna Parma).

Author details

María Piñeiro-Ramil[1,2], Rocío Castro-Viñuelas[1,2], Clara Sanjurjo-Rodríguez[1,2,3], Tamara Hermida-Gómez[2,3], Isaac Fuentes-Boquete[1,2,3], Francisco J. de Toro-Santos[1,3], Francisco J. Blanco-García[2,3]* and Silvia M. Díaz-Prado[1,2,3]

*Address all correspondence to: fblagar@sergas.es

1 Cell Therapy and Regenerative Medicine Group, Department of Biomedical Sciences, Physiotherapy and Medicine, Faculty of Health Sciences, University of A Coruña, Institute of Biomedical Research of A Coruña (INIBIC), University Hospital Complex A Coruña (CHUAC), Galician Health Service (SERGAS), A Coruña, Spain

2 Tisular Bioengineering and Cell Therapy Unit (GBTTC-CHUAC), Rheumatology group, INIBIC, CHUAC, SERGAS, A Coruña, Spain

3 Centro de Investigación Biomédica En Red de Bioingeniería, Biomateriales y Nanomedicina (CIBER-BBN), A Coruña, Spain

References

[1] Blanco-García FJ. Aspectos básicos. Chap. I. In: Batlle-Gualda E, Benito Ruiz P, et al., editors. Manual SER de la artrosis. 1st Ed. Madrid: IM&C; 2002

[2] Bomer N et al. Translating genomics into mechanisms of disease: Osteoarthritis. Best Practice & Research. Clinical Rheumatology. 2015;**29**(6):683-691

[3] Zhang W et al. Current research on pharmacologic and regenerative therapies for osteoarthritis. Bone Research. 2016;**4**:15040-15040

[4] Lories RJ, Luyten FP. The bone-cartilage unit in osteoarthritis. Nature Reviews. Rheumatology. 2011;**7**(1):43-49

[5] Blanco-García FJ, Tornero-Molina J. Artrosis. Chap. 112. In: Farreras P, Rozman C. Medicina Interna. 18th Ed. Barcelona: Elsevier, v.I; 2016.

[6] Kraus VB et al. Call for standardized definitions of osteoarthritis and risk stratification for clinical trials and clinical use. Osteoarthritis and Cartilage. 2015;**23**(8):1233-1241

[7] Allen KD, Golightly YM. Epidemiology of osteoarthritis: State of the evidence. Current Opinion in Rheumatology. 2015;**27**(3):276-283

[8] Ross MH, Pawlina W. Histología. Texto y atlas en color con biología celular y molecular. 5th ed. Madrid: Editorial médica panamericana; 2007

[9] Claus S et al. Cartilage-characteristic matrix reconstruction by sequential addition of soluble factors during expansion of human articular chondrocytes and their cultivation in collagen sponges. Tissue Engineering Part C-Methods. 2012;**18**(2):104-112

[10] Zhang L et al. An in vitro study of collagen hydrogel to induce the chondrogenic differentiation of mesenchymal stem cells. Journal of Biomedical Materials Research Part A. 2012;**100A**(10):2717-2725

[11] Mao JJ. Stem-cell-driven regeneration of synovial joints. Biology of the Cell. 2005; **97**(5):289-301

[12] Gomoll AH et al. The subchondral bone in articular cartilage repair: Current problems in the surgical management. Knee Surgery, Sports Traumatology, Arthroscopy. 2010;**18**(4):434-447

[13] Dhollander AAM et al. The use of scaffolds in the treatment of osteochondral lesions in the knee: Current concepts and future trends. The Journal of Knee Surgery. 2012; **25**(3):179-186

[14] Meretoja VV et al. The effect of hypoxia on the chondrogenic differentiation of co-cultured articular chondrocytes and mesenchymal stem cells in scaffolds. Biomaterials. 2013;**34**(17):4266-4273

[15] Fuentes-Boquete IM et al. Tratamiento de lesiones del cartílago articular con terapia celular. Reumatología Clínica. 2007;**3**:S63-S69

[16] Bhosale AM, Richardson JB. Articular cartilage: Structure, injuries and review of management. British Medical Bulletin. 2008;**87**(1):77-95

[17] Zscharnack M, et al. Repair of chronic osteochondral defects using predifferentiated mesenchymal stem cells in an ovine model. American Journal of Sports Medicine. 2010; **38**(9):1857-69

[18] Magnussen RA et al. Treatment of focal articular cartilage defects in the knee: A systematic review. Clinical Orthopaedics and Related Research. 2008;**466**:952-962

[19] Vilá Y, Rico J, et al. Treatment of osteochondral lesions of the talus with bone marrow stimulation and chitosan-glycerol phosphate/blood implants (BST-CarGel). Arthroscopy Techniques. 2015;**4**(6):e663-e667

[20] Schrock JB et al. Cost-effectiveness analysis of surgical treatment modalities for chondral lesions of the knee: Microfracture, osteochondral autograft transplantation, and autologous chondrocyte implantation. Orthopaedic Journal of Sports Medicine. 2017;**5**(5):2325967117704634

[21] Toh WS et al. Advances in mesenchymal stem cell-based strategies for cartilage repair and regeneration. Stem Cell Reviews and Reports. 2014;**10**(5):686-696

[22] Gomoll AH, Minas T. The quality of healing: Articular cartilage. Wound Repair and Regeneration. 2014;**22**:30-38

[23] Homminga GN et al. Perichondrial grafting for cartilage lesions of the knee. Journal of Bone and Joint Surgery. 1990;**72-B**:1003-1007

[24] Ritsila VA et al. Periosteal and perichondral grafting in reconstructive surgery. Clinical Orthopaedics and Related Research. 1994;**302**:259-265

[25] O'driscoll SW et al. The chondrogenic potential of periosteum decreases with age. Journal of Orthopaedic Research. 2001;**19**:95-103

[26] Díaz Prado SM, et al. Cell therapy and tissular engineering to regenerate articular cartilage. Chap. In: Komorowska MA, Olsztynska-Janus S, editors. Biomedical Engineering, Trends, Researches and Technologies. Ed. Intech, 2011

[27] Valderrábano V et al. Knee-to-ankle mosaicplasty for the treatment of osteochondral lesions of the ankle joint. The American Journal of Sports Medicine. 2009;**37**:105S-111S

[28] Rai V et al. Recent strategies in cartilage repair: A systemic review of the scaffold development and tissue engineering. Journal of Biomedical Materials Research Part A. 2017;**105**(8):2343-2354

[29] Makris EA et al. Repair and tissue engineering techniques for articular cartilage. Nature Reviews. Rheumatology. 2015;**11**(1):21-34

[30] Peterson L et al. Chondrocyte transplantation–An experimental model in rabbits. Transactions of the Annual Meeting of the Orthopaedic Research Society. 1984;**9**:218

[31] Demoor M et al. Cartilage tissue engineering: Molecular control of chondrocyte differentiation for proper cartilage matrix reconstruction. Biochimica et Biophysica Acta-General Subjects. 2014;**1840**(8):2414-2440

[32] Rackwitz L et al. Functional cartilage repair capacity of dedifferentiated, chondrocyte- and mesenchymal stem cell-laden hydrogels in vitro. Osteoarthritis and Cartilage. 2014;**22**(8):1148-1157

[33] Liao CJ et al. Injecting partially digested cartilage fragments into a biphasic scaffold to generate osteochondral composites in a nude mice model. Journal of Biomedical Materials Research Part A. 2007;**81A**(3):567-577

[34] Dewan AK et al. Evolution of autologous chondrocyte repair and comparison to other cartilage repair techniques. BioMed Research International. 2014;**2014**:272481

[35] Litzke LE et al. Repair of extensive articular cartilage defects in horses by autologous chondrocyte transplantation. Annals of Biomedical Engineering. 2004;**32**(1):57-69

[36] Schubert T et al. Long-term effects of chondrospheres on cartilage lesions in an autologous chondrocyte implantation model as investigated in the SCID mouse model. International Journal of Molecular Medicine. 2009;**23**(4):455-460

[37] Meyer U et al. Cartilage defect regeneration by ex vivo engineered autologous microtissue–preliminary results. In Vivo. 2012;**26**(2):251-257

[38] US National Library of Medicine. ClinicalTrials.gov [Online] 2014. http://clinicaltrials.gov/show/NCT01222559.

[39] Dominici M et al. Minimal criteria for defining multipotent mesenchymal stromal cells. The International Society for Cellular Therapy position statement. Cytotherapy. 2006;**8**(4):315-317

[40] Gupta PK et al. Mesenchymal stem cells for cartilage repair in osteoarthritis. Stem Cell Research & Therapy. 2012;**3**(4):25

[41] Baghaban-Eslaminejad M, Malakooty-Poor E. Mesenchymal stem cells as a potent cell source for articular cartilage regeneration. World Journal of Stem Cells. 2014;**6**(3):344-354

[42] Hermida-Gómez T et al. Bone marrow cells immunomagnetically selected for CD271+ antigen promote in vitro the repair of articular cartilage defects. Tissue Engineering Part A. 2011;**17**(7-8):1169-1179

[43] Sanjurjo-Rodríguez C et al. Differentiation of human mesenchymal stromal cells cultured on collagen sponges for cartilage repair. Histology and Histopathology. 2016b;**31**(11):1221-1239

[44] Pak J et al. Potential use of mesenchymal stem cells in human meniscal repair: Current insights. Open Access Journal of Sports Medicine. 2017;**8**:33-38

[45] Muíños-López E et al. Human amniotic mesenchymal stromal cells as favorable source for cartilage repair. Tissue Engineering. Part A. 2017

[46] Kohli N et al. An in vitro comparison of the incorporation, growth, and chondrogenic potential of human bone marrow versus adipose tissue mesenchymal stem cells in clinically relevant cell scaffolds used for cartilage repair. Cartilage. 2015;**6**(4):252-263

[47] Chen FH, Tuan RS. Mesenchymal stem cells in arthritic diseases. Arthritis Research & Therapy. 2008;**10**(5)

[48] Liu Y et al. Therapeutic application of mesenchymal stem cells in bone and joint diseases. Clinical and Experimental Medicine. 2014;**14**(1):13-24

[49] Squillaro T et al. Clinical trials with mesenchymal stem cells: An update. Cell Transplantation. 2016;**25**(5):829-848

[50] Murphy JM et al. Stem cell therapy in a caprine model of osteoarthritis. Arthritis and Rheumatism. 2003;**48**(12):3464-3474

[51] Lee KB et al. Injectable mesenchymal stem cell therapy for large cartilage defects – A porcine model. Stem Cells. 2007;**25**(11):2964-2971

[52] Mokbel AN et al. Homing and reparative effect of intra-articular injection of autologus mesenchymal stem cells in osteoarthritic animal model. BMC Musculoskeletal Disorders. 2011;**12**:259

[53] Sato M et al. Direct transplantation of mesenchymal stem cells into the knee joints of Hartley strain guinea pigs with spontaneous osteoarthritis. Arthritis Research & Therapy. 2012;**14**(1):R31

[54] Koga H et al. Local adherent technique for transplanting mesenchymal stem cells as a potential treatment of cartilage defect. Arthritis Research & Therapy. 2008;**10**(4):R84

[55] Black LL et al. Effect of adipose-derived mesenchymal stem and regenerative cells on lameness in dogs with chronic osteoarthritis of the coxofemoral joints: A randomized, double-blinded, multicenter, controlled trial. Veterinary Therapeutics. 2007;**8**(4):272-284

[56] Im GI et al. Repair of cartilage defect in the rabbit with cultured mesenchymal stem cells from bone marrow. Journal of Bone and Joint Surgery. British Volume (London). 2001;**83**:289-294

[57] Wakitani S et al. Human autologous culture expanded bone marrow mesenchymal cell transplantation for repair of cartilage defects in osteoarthritic knees. Osteoarthritis and Cartilage. 2002;**10**:199-206

[58] Davatchi F et al. Mesenchymal stem cell therapy for knee osteoarthritis. Preliminary report of four patients. International Journal of Rheumatic Diseases. 2011;**14**(2):211-215

[59] Orozco L et al. Treatment of knee osteoarthritis with autologous mesenchymal stem cells: A pilot study. Transplantation. 2013;**95**(12):1535-1541

[60] Wong KL et al. Injectable cultured bone marrow-derived mesenchymal stem cells in varus knees with cartilage defects undergoing high tibial osteotomy: A prospective, randomized controlled clinical trial with 2 years' follow-up. Arthroscopy. 2013;**29**(12):2020-2028

[61] Whitworth DJ, Banks TA. Stem cell therapies for treating osteoarthritis: Prescient or premature? The Veterinary Journal. 2014;**202**(2014):416-424

[62] De Windt TS et al. Allogeneic mesenchymal stem cells stimulate cartilage regeneration and are safe for single-stage cartilage repair in humans upon mixture with recycled autologous Chondrons. Stem Cells. 2017;**35**(1):256-264

[63] Bekkers JE et al. Single-stage cell-based cartilage regeneration using a combination of chondrons and mesenchymal stromal cells: Comparison with microfracture. The American Journal of Sports Medicine. 2013;**41**(9):2158-2166

[64] Toh WS et al. Cartilage repair using hyaluronan hydrogelencapsulated human embryonic stem cell-derived chondrogenic cells. Biomaterials. 2010;**31**(27):6968-6980

[65] Koyama N et al. Human induced pluripotent stem cells differentiated into chondrogenic lineage via generation of mesenchymal progenitor cells. Stem Cells and Development. 2013;**22**(1):102-113

[66] Takahashi K et al. Induction of pluripotent stem cells from adult human fibroblasts by defined factors. Cell. 2007;**131**(5):861-872

[67] Zhao J et al. Induced pluripotent stem cells: Origins, applications, and future perspectives. Journal of Zhejiang University. Science. B. 2013;**14**(12):1059-1069

[68] Tsumaki N et al. iPS cell technologies and cartilage regeneration. Bone. 2015;**70**:48-54

[69] Ko JY et al. In vitro chondrogenesis and in vivo repair of osteochondral defect with human induced pluripotent stem cells. Biomaterials. 2014;**35**(11):3571-3581

[70] Zhang J et al. A human iPSC model of Hutchinson Gilford Progeria reveals vascular smooth muscle and Mesenchymal stem cell defects. Cell Stem Cell. 2011;**8**:31

[71] Liu Y et al. One-step derivation of mesenchymal stem cell (MSC)-like cells from human pluripotent stem cells on a fibrillar collagen coating. PloS One. 2012;**7**

[72] Mimeault M, Batra SK. Recent insights into the molecular mechanisms involved in aging and the malignant transformation of adult stem/progenitor cells and their therapeutic implications. Ageing Research Reviews. 2009;**8**(2):94-112

[73] Yamashita A et al. Generation of scaffoldless hyaline cartilaginous tissue from human iPSCs. Stem Cell Reports. 2015;**4**(3):404-418

[74] Liao J et al. Recent developments in scaffold-guided cartilage tissue regeneration. Journal of Biomedical Nanotechnology. 2014;**10**(10):3085-3104

[75] Kock L et al. Tissue engineering of functional articular cartilage: The current status. Cell and Tissue Research. 2012;**347**(3):613-627

[76] Feng Q et al. Sulfated hyaluronic acid hydrogels with retarded degradation and enhanced growth factor retention promote hMSC chondrogenesis and articular cartilage integrity with reduced hypertrophy. Acta Biomaterialia. 2017;**53**:329-342

[77] He X et al. Layer-by-layer assembly of type I collagen and chondroitin sulfate on aminolyzed PU for potential cartilage tissue engineering application. Applied Surface Science. 2012;**258**(24):9918-9925

[78] Taraballi F et al. Biomimetic collagenous scaffold to tune inflammation by targeting macrophages. Journal of Tissue Engineering. 2016;**7**:2041731415624667

[79] Baino F et al. Bioceramics and scaffolds: A winning combination for tissue engineering. Frontiers in Bioengineering and Biotechnology. 2015;**3**:202

[80] Vinatier C et al. Cartilage engineering: A crucial combination of cells, biomaterials and biofactors. Trends in Biotechnology. 2009;**27**(5):307-314

[81] Quintavalla J et al. Fluorescently labeled mesenchymal stem cells (MSCs) maintain multilineage potential and can be detected following implantation into articular cartilage defects. Biomaterials. 2002;**23**:109-119

[82] Sharma B et al. Human cartilage repair with a photoreactive adhesive-hydrogel composite. Science Translational Medicine. 2013;**5**:167ra6

[83] Benthien JP, Behrens P. The treatment of chondral and osteochondral defects of the knee with autologous matrix-induced chondrogenesis (AMIC): Method description and recent developments. Knee Surgery, Sports Traumatology, Arthroscopy. 2011;**19**(8):1316-1319

[84] Pearce CJ et al. Synthetic osteochondral grafting of ankle osteochondral lesions. Foot and Ankle Surgery. 2012;**18**(2):114-118

[85] Deng Z et al. Cartilage defect treatments: With or without cells? Mesenchymal stem cells or chondrocytes? Traditional or matrix-assisted? A systematic review and meta-analyses. Stem Cells International. 2016;**2016**

[86] Cole BJ et al. Outcomes after a single-stage procedure for cell-based cartilage repair: A prospective clinical safety trial with 2-year follow-up. The American Journal of Sports Medicine. 2011;**39**(6):1170-1179

[87] Benders KE et al. Extracellular matrix scaffolds for cartilage and bone regeneration. Trends in Biotechnology. 2013;**31**(3):169-176

[88] Peretti GM et al. Tissue engineered cartilage integration to live and devitalized cartilage: A study by reflectance mode confocal microscopy and standard histology. Connective Tissue Research. 2006;**47**(4):190-199

[89] Elder BD et al. Extraction techniques for the decellularization of tissue engineered articular cartilage constructs. Biomaterials. 2009;**30**(22):3749-3756

[90] Pei M et al. A review of decellularized stem cell matrix: A novel cell expansion system for cartilage tissue engineering. European Cells & Materials. 2011;**22**:333-343

[91] Barlett W et al. Autologous chondrocyte implantation versus matrix-induced autologous chondrocyte implantation for osteochondral defects of the knee. A prospective randomized study. Journal of Bone and Joint Surgery. British Volume (London). 2005;**87**:640-645

[92] Riboh JC et al. Comparative efficacy of cartilage repair procedures in the knee: A network meta-analysis. Knee Surgery, Sports Traumatology, Arthroscopy. 2016

[93] Wakitani S et al. Repair of rabbit articular surfaces with allograft chondrocytes embedded in collagen gel. Journal of Bone and Joint Surgery. British Volume (London). 1989;**71**:74-80

[94] Sharma S et al. Biomaterials in tooth tissue engineering: A review. Journal of Clinical and Diagnostic Research: JCDR. 2014;**8**(1):309-315

[95] Chiang H et al. Differences between chondrocytes and bone marrow-derived chondrogenic cells. Tissue Engineering Part A. 2011;**17**(23-24):2919-2929

[96] Sanjurjo-Rodríguez C et al. Tissue engineering in an in vitro model of human cartilage repair. Osteoarthritis and Cartilage. 2016a;**24**(Suppl. 1):S169-S170

[97] Nguyen D et al. Cartilage tissue engineering by the 3D bioprinting of iPS cells in a nanocellulose/alginate bioink. Scientific Reports. 2017;**7**(1):658

Section 2

Orthopedics

Chapter 5

Macroscopic Anatomy, Histopathology, and Image Diagnosis of Joints and Synovial Cartilages

Flávio Ribeiro Alves,
Renan Paraguassu de Sá Rodrigues,
Andrezza Braga Soares da Silva,
Gerson Tavares Pessoa, Laecio da Silva Moura,
Jacyara de Jesus Rosa Pereira Alves,
Kássio Vieira Macedo and Robson Giglio

Additional information is available at the end of the chapter

http://dx.doi.org/10.5772/intechopen.70374

Abstract

Joints are physiological connections formed by the association of two or more bones that confer mobility to the skeleton of vertebrates. Composed of several structures, these are often related to pathologies of varied origins, which determine symptomatology of varying degrees of intensity and impairment, responsible for the decrease in life expectancy and the well-being of affected populations. Most of the time, the treatment for these diseases is only symptomatic, aiming at the relief of pain and the return of the patient to daily activities. Thus, there has been an increasing interest in the search for new knowledge about the mechanisms that lead to joint disorders and effective therapeutic resources that may contribute to the fight against pain and to the definitive treatment of joint dysfunctions. To this aim, the knowledge of diagnostic methods, especially imaging methods, is of fundamental importance for the recognition of articular affections, enabling a targeted and effective treatment. Among these auxiliary exams currently used to evaluate the joints, the noninvasive ones are the first choice, where radiography, ultrasonography, magnetic resonance imaging (MRI), computed tomography, and arthroscopy are inserted.

Keywords: diagnostic imaging, arthropathies, technologies, treatments, joint

1. Introduction

Aging populations and rising life expectancy have become a global trend. Developing countries have been living with a growing change in the health profile of the population due to the greater life expectancy. Associated with this, the problems related to chronic degenerative and autoimmune diseases arise, which, if not properly treated and followed over the years, can result in serious health problems, compromising the independence and autonomy of patients affected, especially the elderly. In these countries, chronic diseases have caused important and costly demands on health systems and have interfered in qualitative aspects of life [13].

Noncommunicable chronic diseases and autoimmune diseases are one of the main factors responsible for the decrease in the life expectancy and the well-being of the populations affected. Its prevalence is elevated in elderly patients, where osteoarticular diseases predominate, which account for a significant portion of these [26].

The concept of degenerative osteoarticular disease presupposes hyaline cartilage abnormalities, which determine symptomatology of variable intensity and impairment of function. The clinical picture is called arthrosis, osteoarthrosis, or osteoarthritis. Osteoarthritis is a degenerative condition of articular hyaline cartilage, difficult to diagnose and treat, which affects older patients more frequently, manifested by pain, stiffness, and functional impairment of the affected joint. The degenerative or degradative process of articular cartilage may be primary or secondary to different causes, such as hereditary diseases, endocrine diseases, joint disorders, and inflammatory diseases [28, 53].

Among autoimmune diseases, rheumatoid arthritis, a complex etiology characterized by symmetrical peripheral polyarthritis, which leads to deformity and destruction of the joints due to erosion of bones and cartilage, also presenting a higher incidence in elderly patients stands out. In general, it affects large and small joints in association with systemic manifestations such as stiffness, fatigue, and weight loss. When it involves other organs, the morbidity and severity of the disease are greater and may decrease life expectancy in 5–10 years. With the progression of the disease, the patients develop incapacity to the development of their activities, which generates social and economic impacts [1].

Degenerative joint disease is another arthropathy characterized by a noninflammatory disorder of mobile joints, being considered as a group of disorders defined by the progressive deterioration of articular cartilage, accompanied by bone and soft tissue alterations [11, 59, 63]. This is a chronic condition leading to degeneration of adjacent structures and thickening of the joint capsule. Different factors are identified as the cause of this disease, such as trauma, intra-articular fractures, subluxations or joint dislocations, conformation defects, and angular deformity [37].

Degenerative joint disease manifests initially with mild lameness, which progresses with the development of the disease [34]. In large-moving joints, initial changes are manifested by acute synovitis and capsulitis [56] or muscle atrophies [41], as well as joint capsule distension with an increase in adjacent soft tissue volume [34]. The predominant symptom is pain sensitivity, which may originate from different intra-articular or extra-articular structures, such as capsule, articular cartilage, synovium, periosteum, bones, tendons, bursae, ligaments, or menisci [47].

These data justify an increasing interest in the search for new knowledge about the mechanisms that lead to joint disorders and effective therapeutic resources that can contribute to the fight against pain and to the definitive treatment of joint dysfunctions, preventing the degeneration of structures until irreversible states.

Currently, treatments for these diseases have as main objective the relief of pain and the reduction of functional disability, enabling the development of routine activities and suspension of disease progression. To that aim, several techniques have been proposed, such as pharmacological and nonpharmacological, surgical, and alternative treatments, such as the use of platelet-rich plasma for pain and joint function improvement in osteoarthritis [15, 35], aquatic and nonaquatic exercises [40], and nonsteroidal anti-inflammatory therapies [20], among others.

More recently, cell therapies have been proposed, such as the use of stem cells, which consist of a nonspecialized cell category, that is, they have no tissue-specific structure that allows them to perform individual functions of other cells. These are capable of dividing and renewing themselves over long periods and also of differentiating themselves into specialized cell types. Unlike other cells, such as those of the muscular and nervous tissues, which do not normally replicate, they can replicate several times in a process called proliferation. In this context, the possibility of using stem cells for cellular therapies has become a very coveted area and is the target of several studies, attracting the attention of researchers from all over the world [51].

Primordial germ cell therapies have also been studied for the formation of hyaline articular cartilage due to its regenerative characteristic. Diseases such as traumatic chondral lesions, dissecting osteochondritis, patellar chondromalacia, and osteoarthrosis are targets of therapy with these cells [46].

Traumatic chondral injuries, when moderate and in areas of low mechanical stress, are usually treated by conservative methods that include dietary reduction for weight reduction, analgesics, anti-inflammatories, and physiotherapy. When extensive, more complex treatments are stipulated as autologous or homologous osteochondral grafts, replacement arthroplasty using partial or total prostheses and arthrodesis [17].

The importance of stem cells as a new treatment method in chondral lesions is due to the fact that articular cartilage has little repair capacity. However, the autologous chondrocyte culture transplant technique in chondral defects is still restricted to small lesions and in young patients. In contrast, recent studies have shown that mesenchymal progenitor cells can repair major defects regardless of age. The great difficulty is still the culture, induction of differentiation, and adhesion at the lesion site, which often do not respond as expected [17].

In addition to stem cells, growth factors are also required to determine proliferation and differentiation in cartilaginous tissue both during in vitro cultures and in implantation. These factors include prolactin, which induces cell proliferation and the synthesis of proteoglycans. Other factors that determine chondrogenesis are insulin-like growth factor 1 (IGF-1) and transforming growth factor beta 1 (TGFβ1) [49, 69].

With the advent of this new technique, it is expected that donor area morbidity can be reduced in cases of allografts where small fragments of cartilage are removed from an area of lower load to another with osteochondral defect and reduce contamination and deterioration of these areas, avoiding lesions inherent to more invasive techniques such as release and wear of material, in the cases of joint prostheses [55].

However, the literature has shown in several studies that this topic is one of the most promising fields of medicine, with the potential to provide the resolution of pathologies previously limited to symptomatic treatments [16, 38].

2. Anatomy of the joints

The word articulation originates from the Latin "articulatio" which means rigidity, that is, structure that derives from a cartilaginous bone set of consistent architecture. Physiologically, it is the connection between bones which gives mobility to the skeleton. The joints are formed by the association of two or more bones with the aid of skeletal muscles, ligaments, and joint capsule. The functional activity of the joints depends essentially on the shape of the joint surfaces and the union means, which may limit it [29, 61].

The articular joints are formed by the joint activity of the following structures: bones, articular surface, articular cartilage, joint space, joint capsule, and synovial fluid. Each of these structures plays an important role in the joint [64].

The bones, rigid structures that serve as support and skeleton forming the joints, communicate by favoring the mobility of the body. Depending on the location, the bones may present different anatomical dispositions and therefore infer in the shape and classification of the joints [68].

The articular surfaces are the regions of bone surface that maintain contact for formation of the articular region. These surfaces correspond to the place of insertion of the articular cartilage serving as the base. The latter is the layer of cartilaginous tissue that covers the articular surfaces, absorbing compressive impacts and assisting in the development of the other constituent structures of the joint [39, 68].

The joint capsule is a fibrous sheath that covers the space belonging to the joint while holding the bone structures together. This structure plays the germinative function for the synovial fluid and provides stability to the joints, thus contributing to the creation of an internal portion, of reduced pressure, favoring a better coaptation [8, 29].

Synovial fluid is an aqueous substance secreted by the joint capsule that fills the joint space and ensures lubrication, allowing the stability and distribution of the loads on the surfaces, reducing the stresses of contact. Synovial fluid is a parameter for many articular anomalies, which can be evaluated by means of arthrocentesis (collection of the joint fluid) and by examining the color, appearance, and viscosity of this material [29, 59].

The joints can be classified according to their structure and mobility in fibrous (synarthroses) or immotile movements, cartilaginous (amphiarthroses) or with limited movements, and synovial (diarthroses) or with ample movements. Another type of classification is with regard to the continuity of the bone pieces, which may be continuous (with bone pieces closely connected to each other) and contiguous (where there is a joint cavity) [12].

The fibrous joints, in which the interposed elements between the bony structures are of fibrous nature, called synarthroses (syn: together, arthro: articulation), are immobile joints and can be of three types: sutures, syndesmosis, and gomphosis. The sutures are joints present mainly in the bones of the skull and are characterized by a small amount of fibrous tissue. In syndesmosis, the bone surfaces are joined by a fibrous substance in a tape or ligament aspect that limits the movement of the articular parts, as in the tibiofibular joint. In the gomphosis, the bony structures are irregular, and the pattern is the one of the inserted teeth in their alveoli [12, 32].

Unlike the fibrous ones, in the cartilaginous joints, the interposed tissue is cartilaginous in nature and can be subdivided into synchondrosis and symphysis. Synchondrosis is a provisory or temporary joint, in which the cartilage has a limited life, disappearing soon after the individual reaches adulthood, a situation found in the epiphyseal disks. The symphysis is permanent, commonly present in the intervertebral disks and the pubic symphysis [12, 70].

Unlike fibrous and cartilaginous, the synovial joints allow wide movements, being structurally complex, characterized by the presence of synovial membrane which internally coats the joint space and is responsible for the production of synovial fluid. Other elements participate in the constitution of the synovial joints as the joint cavity, articular bone surfaces, articular cartilage, and articular capsule described previously [32].

Synovial or diarthrosis cartilages are present in most joints and are capable of flexion and extension movements, adduction and abduction, rotation (around the cerebrum-podalic axis, can be medial and lateral), pronation (medial rotation of the forearm), supination (lateral rotation of the forearm), and circumference (joint movement of adduction, flexion, abduction, and extension) [32].

The characteristics found in the articular bone surfaces also allow defining the movements performed by the joint, so these structures can be called flat, seal, ellipsoid, and condylar. The flat surfaces allow sliding movements corresponding to the joints of the carpus or tarsus. In sealing the surfaces that resemble a knight in a saddle, it can be found in the carpometacarpal joint of the thumb. The articulation with ellipsoid surfaces has an elliptical shape, not allowing rotation movements, like the car rim. The condylar, in turn, presents the prominent bone surface appearing a condyle, found in the temporomandibular and metacarpophalangeal [39].

The occurrence of joints involving two distinct natures is possible, as is the case of fibrocartilaginous, which act as shock absorbers, enabling the joint movements. As a way of increasing the contact area of the articular surfaces, the lips (or borders) are examples of joints in which the interposed tissue is fibrocartilaginous in nature. These act as frames and are found in the shoulder joint (glenoid lip). Other examples are disks and menisci. The first, found in the union of the clavicle with the sternum, stabilizes one bony part allowing the other to perform complex movements, as it is also seen in the temporomandibular joint (TMJ). The meniscus, resembling disks, however, is incomplete, acquiring "crescent" form, and is present in the knee joint [66].

Externally, there are elements that reinforce the cohesion between the articular parts, which is the case of the ligaments that can be found internal to the articular or extra-articular cavity and to the physical forces exerted: cohesive force, atmospheric pressure, transition of the coapted bones, and muscular tension [39].

3. Histology of joints

The study of the joints allows inferring about the mechanism of locomotion of the organism, being a content that involves the anatomical part and the ultrastructure of the articular elements. Thus, histology as an important segment in this study defines the tissue characteristics of the joint as well as the importance of its cells for the performance of joint physiology [31].

The component elements of the joints present distinct histological characteristics, where the bone and cartilage tissues are most abundant. The articular surfaces are covered by articular cartilage of the hyaline type. The articular cartilage comprises a highly specialized surface connection fabric that provides a lubricated surface for moving joints and facilitates the transmission and distribution of the loads with a low coefficient of friction [29, 59].

Hyaline cartilage consists of the following cellular elements: chondrocytes, type 2 collagen, and extracellular matrix, as well as important microelements such as water, proteoglycans, glycoproteins, and lipids. Chondrocytes are the most abundant cells in this tissue, which present in their cytoplasm glycogen, lipids, well-developed endoplasmic reticulum, and Golgi complex. These tend to occupy small spaces within the extracellular matrix of the cartilage, called gaps, in which they can be found individually or contain two or more cells by gaps (**Figure 1**) [71].

The hyaline articular cartilage does not present vascularization, and the chondrocytes are nourished by constituents present in the synovial fluid provided by diffusion. The thickness and density of the cartilage vary from joint, and in humans, it is thicker on the end of the femur

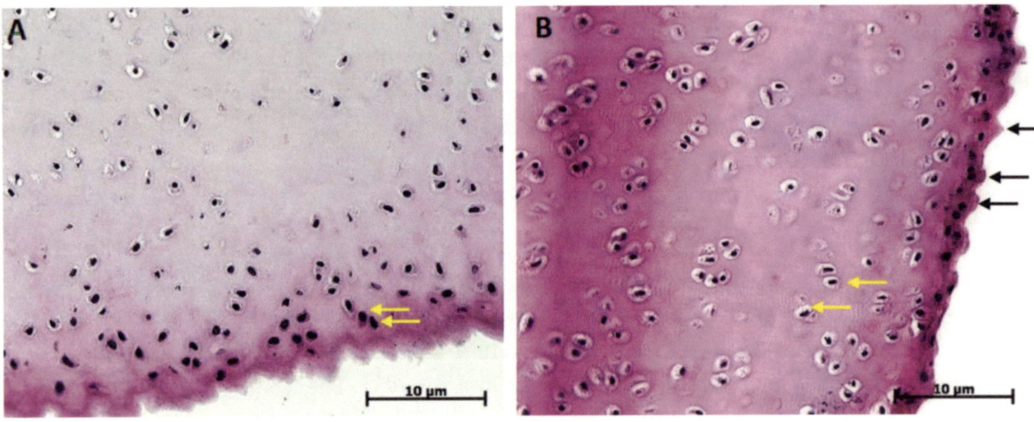

Figure 1. Photomicrograph of hyaline cartilage from a CAE model (caprine arthritis and encephalitis model) of an infected goat. (A) Affected SHJ (humerus head surface). Note the irregular joint surface with loss of cartilage integrity and heterogeneous chondrocyte distribution that are seen flattened on every surface aspect (arrows) and focal degeneration with cartilage fibrillation (wide arrows). (B) Carpal joint (carpal radial bone). Note the irregular perichondrium surface with spaced and little evident chondroblasts. The chondrocytes wrapped in matrix (*) are also seen in fewer quantities and spaces on the surface and deep layers. Bars: (a) 10 μm and (b) 10 μm (image gentile provided by Professor Flavio Alves, Specialized Veterinary Diagnostic Imaging Laboratory (LABDIVE), Federal University of Piauí, Teresina, Piauí, Brazil).

and the tibia, ranging from 2 to 4 mm. From this thickness, four distinct layers are divided according to the cellular morphology and structure of the extracellular matrix in a superficial, transient, deep, and calcified cartilage zone. The arrangement of chondrocytes and collagen fibers varies between layers, increasing cell density as it approaches the articular surface [62].

The superficial or tangential layer is responsible for the slip of the movement of the bony parts and lubrication, composing about 20–30% of the articular cartilage. This zone is composed of two layers, a thin fibrillar lamina without cells (located in the bed more superficial or distal to the articular surface) and another layer of flat chondrocytes and collagen fibers oriented tangentially to the articular surface, having low proteoglycan content [62].

The transitional or intermediate layer is responsible for the transition between the shear forces of the articular parts, still corresponding to about 60–70% of the cartilage; this layer is composed of relatively larger round chondrocytes and immersed in an extracellular matrix. In this area, the collagen fibers are thick and randomly arranged, with a high content of proteoglycan with the presence of spherical chondrocytes. Finally, the calcified or deep layer establishes an intimate relation with the articular surface, corresponding to the smaller percentage in the constitution of the cartilage [52, 72].

The cartilaginous matrix is constantly subjected to external forces due to movement and the load imposed on the joints, which impose the need to maintain high resistance and flexibility. These characteristics are conferred by the collagen fibrils and the amorphous intercellular substance, which are inserted in their constitution permeated by a collagen network composed of water, proteoglycans, and hyaluronic acid. Water is the most abundant element in the matrix, and its high content in the cartilage favors the absorption of impacts, giving the articular cartilage the deformity necessary to withstand the compressive forces to which it is normally subjected. In addition, the cartilage matrix contains electrolytes such as Ca^{2+}, Na^{+}, and K, in concentrations higher than those found in synovial fluid [67].

Chondrocytes are the main cellular elements found in the articular cartilage and produce different collagen molecules, type II collagen being the most abundant in the joints. This collagen is characterized by three $\alpha 1$ chains of type II and organized in fibrils that give a three-dimensional network shape to the matrix allowing a certain degree of deformity when it is subjected to compressive or tensile forces [31].

In addition to water and hyaluronic acid, the matrix consists of proteoglycans, complex molecules composed of glycosaminoglycans, which are polysaccharides made up of sulfated disaccharide units that repeat themselves in relatively short and unbranched chains. The proteoglycans bind to hyaluronic acid forming chains of multi-molecules favoring the cellular organization of the matrix [31].

When synthesized and secreted by the chondrocytes, the hyaluronate-proteoglycan complexes and the collagen cluster themselves, resulting in perfectly structured complexes adapted to withstand the compression and traction forces to which the joint is subjected. Once the cartilage is subjected to compressive forces, the water retained by the proteoglycans is released proportionally to the force exerted, being recovered when that force is ceased. However, the amount of water that proteoglycans can expel upon being compressed is limited and determined by their charge [68, 72]. Thus, the ability of articular cartilage to withstand compressive forces is directly proportional to the concentration of proteoglycans in the matrix and depends on the maintenance of its integrity, which at times may subject it to ruptures [29].

In addition to the articular cartilage, other elements are involved in the ultrastructure of the joints as the synovial fluid and the joint capsule. The intra-articular space, located between two opposite bone ends, contains the synovial fluid, which lubricates the articular surfaces, reducing friction, and serves as a vehicle for the diffusion of nutrients from the blood vessels of the synovial membrane to the articular cartilage chondrocytes. The elimination of the end products of the cellular metabolism occurs through mechanisms of diffusion, through the cartilage, to the blood and lymphatic vessels of the bone and the synovial membrane [30].

The synovial membrane that coats the articular capsule internally lies close to the surface of the cartilage, separated only by the synovial fluid, and is composed of two leaflets: the first (internal) is the synovial intima, devoid of basement membrane, and composed of one to four layers of cells. The second (more external) connects the outer wall of the fibrous capsule with the synovial intima, which is formed by loose connective tissue with fenestrated capillaries [5, 31].

The synovial intima is composed of two cell types: the "A"-type cells, similar to macrophages (because they have the same derivation of monocytic cells from the bone marrow), and the "B"-type cells, called synoviocytes, which have characteristic fibroblasts. This membrane covering the synovial fluid functions as a dialysis membrane, which, due to the increased capillary hydrostatic pressure, allows the ultrafiltration of the blood, the synovial fluid being constituted by the ultrafiltrate that passes from the synovial capillaries to the joint cavity. The articulation presents microelements essential for its activity in the midst of external and internal compressive forces, as well as assisting in the renewal and integrity of the tissues that compose them [31].

4. Diagnostic methods in articulation

Often, the joints are affected by inflammatory, infectious, or degenerative conditions that can reach the cartilage, bones, and adjacent structures or a combination of these, causing serious damage to the patient. The treatment of these pathologies is elaborated through the definitive diagnosis, which usually relies on the accomplishment of complementary exams, especially the imaging [57].

Imaging methods are essential in the diagnosis of bone and joint changes. Among the auxiliary examinations currently used to evaluate the joints, the noninvasive ones are the first ones of choice, where radiography, ultrasonography, magnetic resonance imaging, computed tomography, and arthroscopy are inserted. Usually, the evaluation begins with the radiological examination, capable of providing essential information about the bony and articular cartilaginous structures. The imaging tests are used to evaluate the integrity of the articular components and the relationship between them, confirm the extent or stage of disease progression, and evaluate the effects of the treatments performed [57].

4.1. Radiography

Radiography is the most common imaging technique, based on imaging by X-ray transmission over a target tissue. The rays that go beyond the body reach a film, sensitizing it. After the revelation, the rays that are absorbed in the body do not sensitize the film, and the corresponding areas will be white (radiopaque). On the other hand, the sensitized areas make the regions in the film black (radiolucent). In the analysis of the film, a variation of shades from white to black denominated radiological density is observed. The contrast between the light and dark areas in the radiography depends on the technical and physical conditions in the capture of the images [10].

Like other techniques that expose the body to radiation, X-rays are harmful, requiring the adoption of procedures aimed at protecting exposed professionals and patients. The damage caused by ionizing radiation is cumulative, which means that the harm is caused by repeated doses of radiation that accumulate in the tissues. In order to minimize these risks, collimators, radiation dosage control, plumb protection, screens, and individual monitors (dosimeter) are used for professionals who deal daily with this type of examination [22].

After the technical adjustments and taking into account the biosafety tools, the region to be analyzed in the radiography must be properly positioned so that favorable images are acquired for its evaluation. Thus, it is fundamental that incidences are made in different positions, determining opposite and/or complementary planes [6].

In the attempt to improve differentiation between structures of similar density, such as those found in the abdomen, contrast media are used which may be either natural (air) or artificial (barium based and iodine based). These solutions are mainly used in the study of digestive, urinary, biliary, vascular, and joint studies [10].

Radiography is an important diagnostic method for the study of joint changes. However, fractures in rigid structures, neoplasias, growth and posture disorders, traumatic and inflammatory changes, deposition of substances, and problems of calcification, among others, can be diagnosed. It has a high interest in the evaluation of the progression of rheumatic diseases and in the diagnosis of their complications. The radiographic changes found will vary according to the type of lesion and the time of evolution, keeping the clinician informed about the severity of the condition [24].

The radiographic analysis of the joints should take into account the joint space, its dimensions, and regularities. The thickness of the joint space consists of the joint dimension of the cartilages of both bone structures. Any interference in this space can be represented in the radiographic image and indicate inflammatory changes as in the cases of arthritis. The space may be diminished in the case of advanced arthropathies, which may be asymmetric or localized, depending on the pathology, or it may occur that a loss of space is generalized [65].

Synovium, synovial fluid, and articular capsule, because they have the same radiodensity as adjacent soft tissues and cartilage, are only seen if they are contoured by a radiant layer. For this reason, it is often necessary to complement the simple X-ray with the use of articular contrast media, known as arthrography (**Figure 2**) [10, 48].

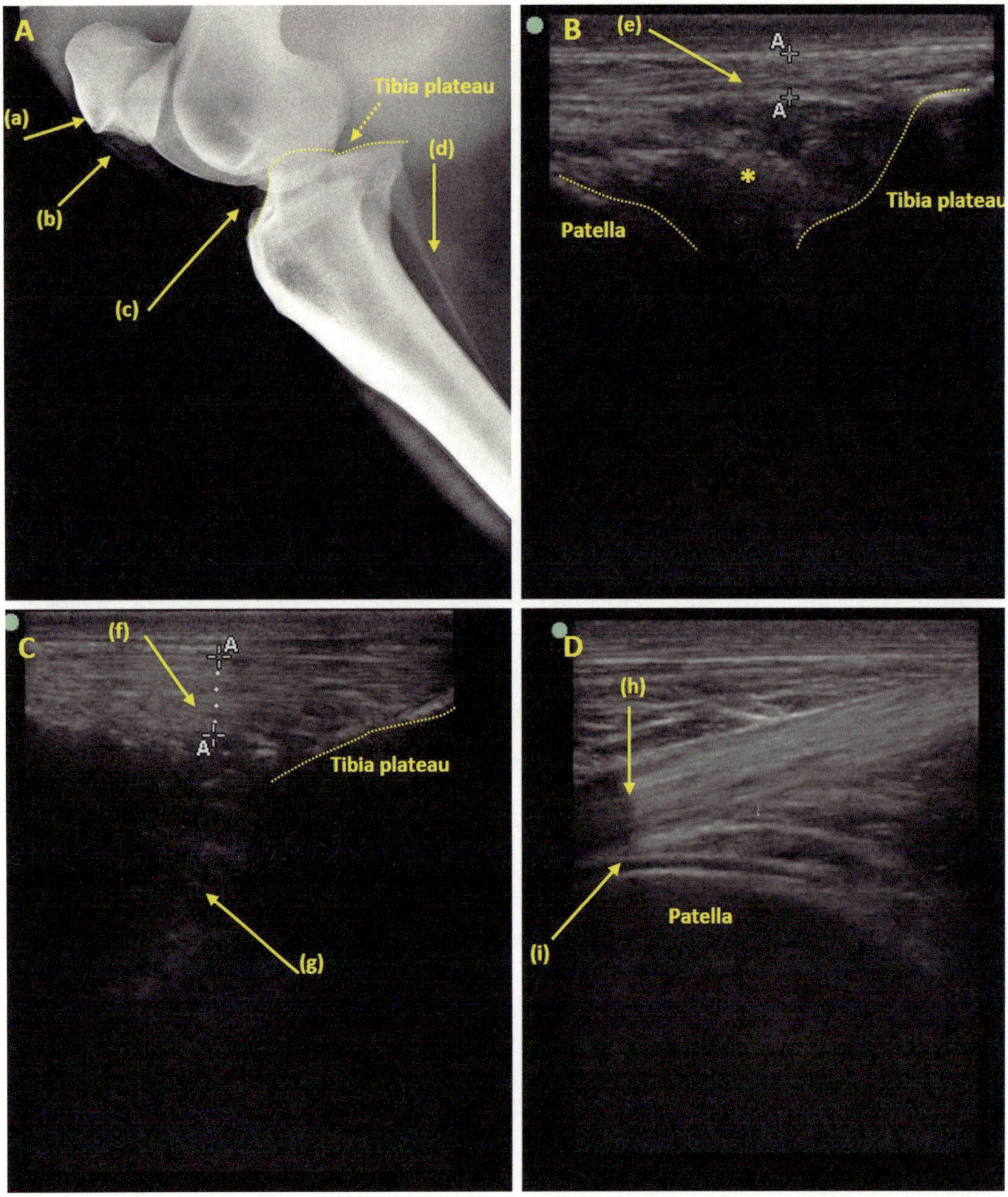

Figure 2. Radiographic and ultrasound imaging of a normal equine knee joint. (A) Note the smooth surface of the joint (femoral head and tibial plateau), with the discreet presence of the patellar ligament (b), due to the high incidence of X-ray bundles. (B and C) The normal ultrasonographic pattern of the patellar ligament, showing homogeneous echotexture and habitual echogenicity. Note the parallel arrangement of the tendon fibers and the normal hyperechogenic appearance of the infrapatellar fat pad (*). (D) Proximal insertion of the patellar tendon (h). (a) Patella, (b) patellar ligament, (c) joint space, (d) fibula, (e and f) patellar ligament echotexture, (g) infrapatellar fat pad, (h) proximal insertion of the patellar tendon, and (i) joint space. (Image gentile provided by the Diagnostic Imaging Services, Federal University of Piauí, Teresina, Brazil).

The arthrography corresponds to the contrasted representation of the joint space, and the viability of using the technique with a positive (iodized) contrast is injected directly into the joint. Unlike the simple radiography, the arthrography should be performed with the patient in sedation due to the discomfort in the application of the contrasts. This technique is performed to demonstrate and assess arthropathies and associated soft tissue structures [36, 65].

There are indications of arthrography when there is suspicion of soft tissue ruptures present in the joint space, which are not adequately visualized in the simple radiography, due to the minimal differentiation of radiological density. However, many contrasts may trigger undesirable reactions, so this technique is infeasible in case of patients allergic to contrast or solutions used in sedation [10].

Currently, double-contrast arthrography in the joints has been used in humans both in radiology and associated to computed tomography, in order to identify lesions on joint surfaces and in nonbone structures, which has shown great advantages when compared to arthrography with positive contrast medium [50].

4.2. Ultrasonography

Ultrasonography presents as a consolidated and sensitive examination for the observation of periarticular soft tissue alterations of the articular surfaces, besides being able to diagnose the morphological changes promoted by various arthropathies early [60]. This is due, in large part, to the improvement in the image quality of the equipment, due to the improvement of the imaging technology and the manufacture of transducers with increasing resolution, in addition to the relative decrease in the price of the equipment (**Figure 2**) [21].

This technique presents some advantages compared to the radiography because it is a noninvasive examination, able to detect early changes, besides providing details of the tissue parenchyma and evidencing structures that do not appreciate the radiographic examination [21].

Such information can be seen by means of the changes that occur in the synovial membrane, joint capsule, as well as periarticular volume increase. This technique allows direct visualization of the joint space, besides being able to guide needles in real time, in cases of treatments with intra-articular drug infusions. Furthermore, it can guide treatments according to signs of inflammation and allows the visualization of the appropriate distribution of medication within the joint space [7, 58].

In general, it is not necessary to pre-prepare the patient for ultrasonographic joint examination, only the application of a thick layer of acoustic gel between the transducer and the ultrasound window to reduce the interference of the layer of air on the skin [21].

Lately, ultrasound examination has been gaining space as a complementary diagnostic method in the therapeutic follow-up of several joint diseases such as rheumatoid arthritis, synovitis, bone erosions, mainly psoriatic arthritis, and systemic lupus erythematosus. The great advantage of the sonographic study is its ability to detect changes such as synoves and bone erosion early on radiography, which has been increasingly valued in the prevention of late and definitive structural damage [3].

Depending on the frequency used in the transducers, it is possible to evaluate most joints by means of ultrasonography. With it, one can investigate structures such as tendons, brackets, cartilage, and bone surface, making it possible to search for erosions in inflammatory diseases in general. The possibility of evaluating numerous structures in a single study extends its application in several rheumatologic pathologies, such as rheumatoid arthritis, spondylarthritis, arthritis by microcrystals, osteoarthritis, collagenosis, and systemic vasculitis. The use of ultrasound is effective for the determination of the presence or absence of lesions in tendons and should be considered as a first line of diagnostic tool [25].

In articulations, ultrasonography is used to evaluate the response to treatment, aiming to reduce the degree of synovitis by examining gray scales and/or synovial vascularization using the Doppler technique in its various modalities. Several ultrasonographic degrees of synovial involvement are proposed in the literature, which have as main objective the detection of possible alteration of the inflammatory activity, analyzing the smallest number of joints possible, to reduce the time of the exam execution [4].

Ultrasonography has a good correlation with magnetic resonance imaging (MRI) in the detection of synovitis and erosions. However, although MRI is considered the gold standard for detection of joint changes, this examination is often uncomfortable for patients besides being contraindicated in the holders of metallic prostheses due to the possibility of physical damages. Also, it is a time-consuming, expensive exam that requires the use of a contrast medium, making evaluation of many joints in a single moment impossible. Thus, ultrasound has assumed an important advantage as a highly feasible method in the diagnostic and sequential treatment of patients with various arthropathies. This can be done more frequently, allowing the evaluation of the progress of the treatment and allowing real-time and dynamic analysis, with the joint in motion.

Recent studies with ultrasound of the ankle joint in patients with Chikungunya, despite the limitations of this study, have made possible the characterization and quantification of the sonographic alterations related to this disease, highlighting the role that the method plays in the diagnosis of such complications. The predominant findings in this study were effusion and tenosynovitis, mainly fibular and posterior tibial, and the most common musculoskeletal comorbidity was the involvement of the calcaneus tendon [44].

4.3. Arthroscopy

Although arthroscopy is a surgical procedure, it is a minimally invasive technique, with a relatively fast execution and good postsurgical recovery, allowing the observation of the interior of a joint through the use of a device called an arthroscope. The arthroscope is an endoscope-like apparatus, consisting of a thin rigid cylindrical tube, the thickness of a pencil, containing a microcamera coupled to the end, carrying optical fibers, which transmit images to a TV monitor, allowing the visualization of the inner face of the joint. The evaluation of the articular surface through arthroscopy solves the limitations of the traditional methods of the examinations like the radiography and ultrasonography, allowing the precise diagnosis of articular alterations [9].

With the development of this technique, associated with the discovery of predisposing factors to various arthropathies, restoration of function through minimally invasive procedures, essentially eliminating lesions and helping patients return to normal activities, was even more safe and effective [18].

Arthroscopy is indicated for the diagnosis of joint affections, for the follow-up of treatments and evolution of diseases and in cases of intra-articular alterations not diagnosed by conventional imaging techniques. Arthroscopy of hip-like joints offers minimally invasive surgery for procedures that would require hip dislocation, a more complicated technique. In this joint, the most commonly treated pathologies are femoroacetabular impacts, which are closely associated with demanding activities in hip flexion and internal rotation, common in sports such as golf, baseball, ice hockey, and soccer [7, 18, 43, 54].

Diagnostic indications involve the evaluation of cartilage in osteonecrosis or in conjunction with osteotomies and painful arthroplasties and the collection of tissues for culture. Moreover, synovial diseases such as chondromatosis, pigmented villonodular synovitis, and rheumatoid arthritis are a good indication for this procedure, as well as the treatment of deep gluteal pain [9].

New indications for arthroscopy are being tested, such as round ligament reconstruction, capsulorrhaphy in cases of instability, and repair of tendinous lesions. It is not recommended, however, in cases where there is an infectious process installed in the joint or active skin infections, except when this procedure has the objective of draining secretions resulting from septic arthritis or evaluation of the degree of infection in prostheses [9].

In general, the preparation for the arthroscopy exam is similar to any other surgical procedure. The physician should have all clinical data on the patient as well as information on hypersensitivity reactions to any medication, including anesthetics, the use of medications, associated health problems, vascular problem such as thrombosis or bleeding, and the possibility of gestation. In addition, general and specific preoperative examinations should be performed for a safer procedure [14].

The procedure is performed with the anesthetized patient, which will depend on the structure to be manipulated, ranging from epidural or spinal anesthesia, for procedures in the pelvic limbs, to general anesthesia for shoulder or hip interventions. Sedative drugs are usually given, and the patient sleeps during the examination, however, can be performed with the patient awake. The patient remains monitored by the anesthesiologist until the end of the procedure, being evaluated the parameters such as heart rate, blood pressure, respiration, body temperature, and cardiac electrical activity, among others [14].

For the realization of the technique, two small accesses are realized in the articulation: the first one where the arthroscope will be introduced and the second to direct the necessary instruments for the operation, if necessary. In general, a certain amount of saline is inserted into the joint so that it is inflated and becomes clearer, thus allowing a better visualization. Also, tourniquets can be performed to temporarily reduce blood flow, which could hamper visualization. Thus, therapeutic procedures such as removal, reconstruction or repair of menisci or ligaments, removal of loose bone fragments, or cartilage within a joint or inflamed synovial tissue are possible [14].

Studies with high-performance soccer athletes have shown that hip arthroscopy for the assessment of pathologies of this joint, such as the femoral acetabular impact (FAI), has been shown to be a safe procedure with satisfactory results regarding the return of the athlete to sporty activities. Hip arthroscopy in athletes with symptomatic FAI and labral pathology allowed for complete rehabilitation, earlier than those undergoing open surgery.

Hip arthroscopy is a safe treatment method for a majority of hip pathologies that were unknown until the last decade. The instruments and surgical technique of hip arthroscopy continue to evolve. Better and better results and fewer complications should be expected according to the learning curve.

4.4. Magnetic resonance imaging

Discovered in 1946 by researchers at Stanford University, magnetic resonance imaging (MRI) has been implanted in medicine by Purcell at Harvard years later. In medicine, the first images were obtained from 1972 and advances provided by the application of the technique provided the nomination of Paul Lauterbur and Peter Mansfield to the Nobel Prize of Medicine. In Brazil, the technique was first implanted in the Albert Einstein Hospital of São Paulo in 1986 [27].

MRI is a diagnostic imaging method that uses a magnetic field and radiofrequency waves to obtain images of the interior of the objects in the form of tomes or cuts, without the availability of ionizing radiation. For this, it is necessary to understand physical principles related to the acquisition of images, among them, subjects about electromagnetism, superconductivity, and signal processing [19, 27]. In the clinical setting, MRI aims to complement the diagnostic conclusion given by conventional imaging tests [42].

The formation of the MR image is the result of the interaction of the strong magnetic field produced by the equipment with the hydrogen protons of the living tissue, formulating a condition so that a pulse of radiofrequency can be sent and after collecting the differentiated radiofrequency through a receiving instrument. The signal encoded due to a magnetic field gradient is collected, processed, and converted into an image or information [42].

Hydrogen is the chemical element with the highest concentration in the tissues and with the greatest magnetic moment (the capacity to produce the highest radio signal of all the stable nuclei). Therefore, it is used as the signal source in most magnetic resonance imaging tests. Once a tissue is subjected to a magnetic field and left long enough, the tissue magnetization (name given to the process of interaction of the equipment with the hydrogen protons of the tissue) reaches an equilibrium value that is proportional in intensity to the external magnetic field [45].

Some organs produce a stronger or weaker signal than others, going according to the density of hydrogen present in that tissue, for example, adipose tissue, cerebrospinal fluid, blood, and other body fluids that produce a strong signal due to high density of protons. In contrast, in the absence or low density of mobile protons in the tissue, there will be a zero or very small value capable of overriding the evaluation parameters at resonance [27, 42].

All soft bodies can be seen in MRI; however, the cortical bone and air do not produce signal in the images because of the inability of the protons to relax in the dense bone matrix and the relative lack of hydrogen nuclei in the air. Thus, due to the low density of mobile protons, the lenses do not show any signal in any sequence used. All other structures are visible in varying degrees from gray to white because of variations in signal strength. This differentiation between proton densities in tissues defines, in medical terms, the occurrence of tissue changes, as it increases the difference between a lesion and a surrounding tissue [27, 42].

In general, MR imaging is based on the relationship between the equipment and the living tissue so that the patient's atomic nuclei align along the applied magnetic field, generating a magnetization vector. Subsequently, sequential magnetic field gradients are applied to the spatial location of the signals to be acquired; thus, the excitation pulses are applied, and the nuclei absorb energy. After the excitation pulses are applied, the relaxation phenomena begin, and the nuclei begin to induce the MRI signal in the receiver coils. This signal is acquired and processed by means of the transformed Fourier, where the image is formed point to point in a matrix [2].

However, for the execution of the examination, the anatomical and clinical prior knowledge of the radiologist technician is still necessary. In the sequence, it is of great value to obtain the best images, as well as to minimize artifacts of techniques. Choosing the appropriate coil for the study region that provides a better signal for exam quality and proper patient positioning are imperative items in the MRI [23].

According to the indication, specific protocols are established for the region to be examined and can be divided into the regions: central nervous system, thorax, abdomen, pelvis, and musculoskeletal system. In general, it is indicated that the patient is placed in dorsal decubitus with the head resting on the appropriate coil (quadrature) with the region of interest straight and in the center of the magnet, upper limbs extended on the side of the body and support for the legs in order to promote alignment of column curvatures [2].

In order to evaluate joints, magnetic resonance imaging becomes an excellent diagnostic modality, since it allows identification of not only bone and cartilage structures but also soft tissues such as meniscus, ligaments, cortical and medullary bone compartment, muscles, tendons, and fat (**Figure 3**) [33].

It is believed that the greatest advantage of this technique for joint evaluation is the detection of the disease by the investigation of alterations in the articular components, such as the thickening and enhancement of the synovial membrane, a situation found in rheumatoid arthritis and easily demonstrated by the intravenous injection of paramagnetic contrast (gadolinium). In addition, MRI stands out as a noninvasive method, useful as a complement to clinical

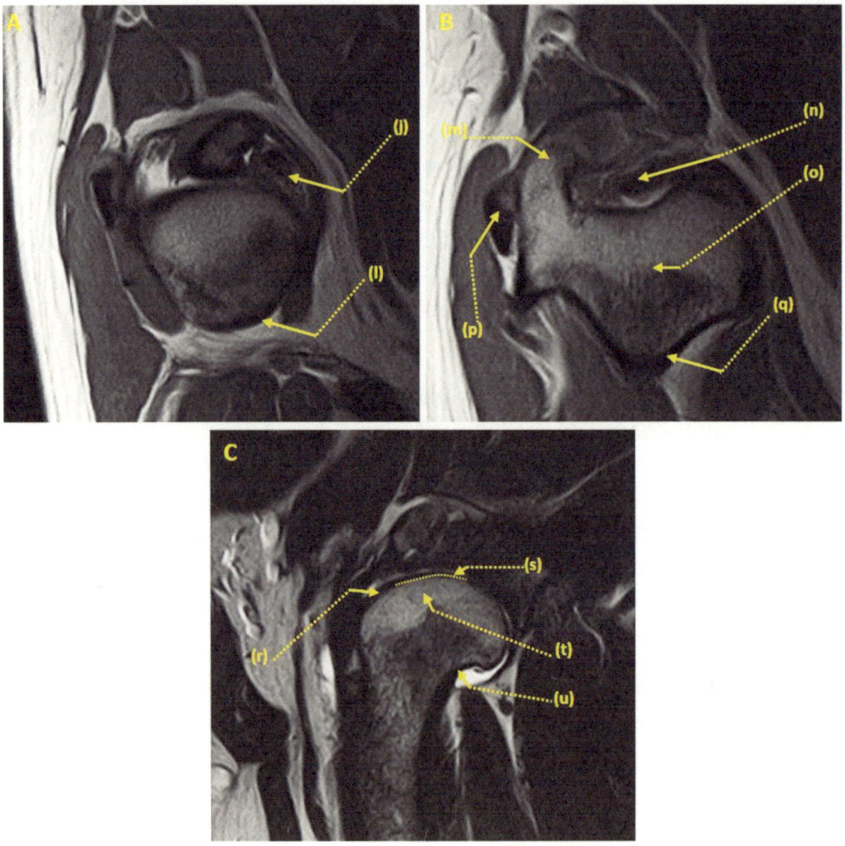

Figure 3. (A–C) Magnetic resonance of a normal canine shoulder joint. (j) Subscapularis tendon, (l) joint space, (m) greater tubercle, (n) biceps tendon, (o) humeral head, (p) supraspinatus tendon, (q and r) cranial joint space, (s) cartilage surface, (t) subchondral bone, and (u) caudal joint space (image gentile provided by Professor Robson Giglio, Department of Small Animal Clinical Sciences, College of Veterinary Medicine, University of Florida).

joint assessment, not only for detection of early disease changes but also for its evolutionary control, treatment monitoring, and differential diagnosis with other diseases (**Figure 4**) [27].

In addition, MRI allows measurement of the extent of joint and extra-articular involvement and evaluation of complications due to disease time, with a higher sensitivity for the evaluation of tendon and ligament injuries, involvement of the tendon sheath (tenosynovitis), trochanteric pouch, bone lesions (subchondral erosions, cysts) that initially may not be seen by conventional radiography, changes in bone marrow, chondral lesions, and in the differentiation between joint effusion and synovitis, using paramagnetic contrast that does not pose risks to the patient (**Figure 5**) [42].

However, in spite of the high cost and its limitations for its execution, magnetic resonance imaging in general still constitutes the best imaging method for joint evaluation, standing out for the other examinations due to its advantages of noninvasiveness, the absence of ionizing radiation, not the use of iodinated contrast (potentially nephrotoxic and allergenic), and ability to better anatomical detail, both by the multiplanar nature of acquisition and by the high contrast between different body tissues [27].

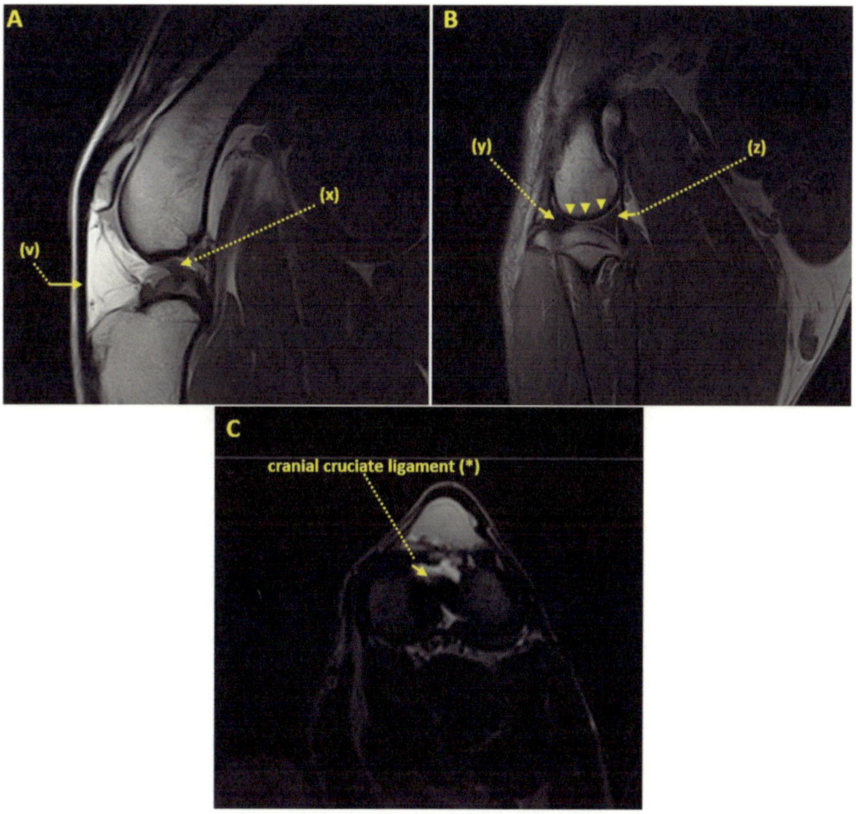

Figure 4. (A–C) Magnetic resonance of a normal canine knee joint. (v) Patellar ligament, (x) cranial cruciate ligament, (y and z) meniscus, joint surface (arrowhead), and (*) cranial cruciate ligament (image gentile provided by Professor Robson Giglio, Department of Small Animal Clinical Sciences, College of Veterinary Medicine, University of Florida).

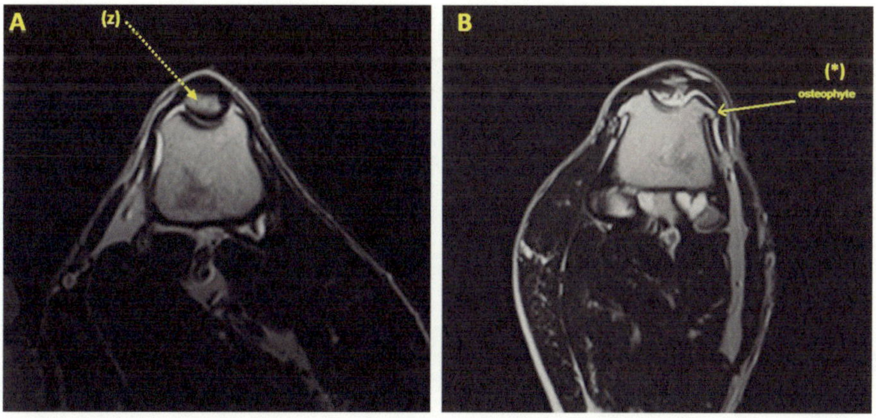

Figure 5. (A and B) Magnetic resonance of a normal canine shoulder joint *versus* osteoarthrosis. (z) Biceps tendon and (*) osteophyte. Note the reduction of joint space and discrete synovial edema, associated with irregularity of articular cartilage (image gentile provided by Professor Robson Giglio, Department of Small Animal Clinical Sciences, College of Veterinary Medicine, University of Florida).

5. Conclusion

Advances in technologies related to research on the diagnosis and treatment of joint diseases have demonstrated excellent results, contributing to the quality of life of patients affected and their return to daily activities. The improvement in the quality of the imaging equipment, combined with the various works in the area of rheumatology, has contributed to a better clinical management of patients, allowing a more conclusive diagnosis and, consequently, the implementation of effective treatments.

Acknowledgements

Our thanks go to Professor Robson Giglio of the University of Florida for the granting of illustration images of magnetic resonance. In addition, we thank the Diagnostic Imaging Services of the University Veterinary Hospital of the Federal University of Piauí (UFPI) for the concession of the images of articular ultrasonography.

Author details

Flávio Ribeiro Alves[1*], Renan Paraguassu de Sá Rodrigues[1], Andrezza Braga Soares da Silva[1], Gerson Tavares Pessoa[2], Laecio da Silva Moura[1], Jacyara de Jesus Rosa Pereira Alves[3], Kássio Vieira Macedo[4] and Robson Giglio[5]

*Address all correspondence to: flavioribeiro@ufpi.edu.br

1 Department of Morphophysiology, Federal University of Piauí, Teresina, Brazil

2 Veterinary Diagnostic Imaging Residency, Veterinary Hospital, Federal University of Piauí, Teresina, Brazil

3 Coloproctology and Colorectal Surgery Service of the University Hospital, Federal University of Piauí, Teresina, Piauí, Brazil

4 Postgraduate Dentistry, Federal University of Piauí, Teresina, Brazil

5 Department of Small Animal Clinical Sciences, College of Veterinary Medicine, University of Florida, Gainesville, Florida, USA

References

[1] Albers JM, Paimela L, Kurki P, Eberhardt K, Emery P, Hof M, Avan't Schreuder F, Leirisalo-repo M, Van Riel PLCM. Treatment strategy, disease activity, and outcome in four cohorts of patients with early rheumatoid arthritis. Annals of the Rheumatic Diseases. 2001;**60**:453-458

[2] Amaro Junior E, Yamashita H. Aspectos básicos de tomografia computadorizada e ressonância magnética. Revista Brasileira de Psiquiatria. 2001;**23**:2-3. DOI: 10.1590/S1516-44462001000500002

[3] Arend CF. Ultrassonografia em portadores de artrite reumatoide: o que o reumatologista clínico deve saber. Revista Brasileira de Reumatologia. 2013;**53**(1):1-6. DOI: 10.1590/S0482-50042013000100009

[4] Backhaus M, Burmester GR, Sandrock D, Loreck D, Hess D, Scholz A, Blind S, Hamm B, Bollow M. Prospective two year follow up study comparing novel and conventional imaging procedures in patients with arthritic finger joints. Annals of the Rheumatic Diseases. 2002;**61**(10):895-904. DOI: 10.1136/ard.61.10.895

[5] Bari C, Dell'Accio F, Tylzanowski P, Luyten FP. Multipotent mesenchymal stem cells from adult human synovial membrane. Arthritis and Rheumatism. 2001;**8**(44):1928-1942. DOI: 10.1002/1529-0131(200108)44:8<1928:AID-ART331>3.0.CO;2-P

[6] Bontrager KL, Lampignano JP. Tratado de posicionamento radiográfico e anatomia associada. Brasil: Elsevier; 2005

[7] Bruyn G, Schmidt W. How to perform ultrasound-guided injections. Best Practice & Research. Clinical Rheumatology. 2009;**23**(1):269-279. DOI: 10.1016/j.berh.2008.11.001

[8] Buckwalter JA, Mankin HJ, Grodzinsky AJ. Articular cartilage and osteoarthritis. Instructional Course Lectures-American Academy of Orthopaedic Surgeons. 2005;**54**:465. DOI: 10.1007/s11420-011-9250-z

[9] Byrd JW, Jones KS. Artroscopia de quadril em atletas: seguimento de 10 anos. American Journal of Sports Medicine. 2009;**37**(11):2140-2143. DOI: 10.1590/S0102-36162009000100004

[10] Canevaro L. Aspectos físicos e técnicos da radiologia intervencionista. Revista Brasileira de Física Médica. 2009;**1**(3):101-115

[11] Caron JP. Osteoarthritis. In: Roos MW, Dyson SJ, editors. Diagnosis and Management of Lameses in the Horse. Philadelphia: Saunders Company; 2003. p. 594

[12] Carrere MTA. Biomecánica clínica. Biomecánica articular. REDUCA (Enfermería, Fisioterapia y Podología). 2010;**3**(2):14-31

[13] Chaimowicz F. A saúde dos idosos brasileiros às vésperas do século XXI: Problemas, projeções e alternativas. Revista de Saúde Pública. 1997;**31**:184-200. DOI: 10.1590/S0034-89101997000200014

[14] Chokshi BV, Rosen JE. Diagnostic arthroscopy of the knee. In: Koval KJ, Zuckerman JD. Atlas of Orthopedic Surgery: A Multimedia Reference. Lippincott Williams and Wilkins: Philadelphia. 2004. 554 p

[15] Coimbra IB, Pastor EH, Greve JMD, Puccinelli MLC, Fuller R, Cavalcanti FS, Maciel FMB, Honda E. Osteoartrite (artrose): tratamento. Revista Brasileira de Reumatologia. 2004;**6**(44):450-453. DOI: 10.1590/S0482-50042004000600009

[16] Cristante AF, Barros-filho TE, Tatsui N, Mendrone A, Caldas JG, Camargo A, Alexandre A, Teixeira WG, Oliveira RP, Marcon RM. Stem cells in the treatment of chronic spinal cord injury: Evaluation of somatosensitive evoked potentials in 39 patients. Spinal Cord. 2009;**47**(10):733-738. DOI: 10.1038/sc.2009.24

[17] Cristante FA, Narazaki DK. Avanços no uso de células-tronco em ortopedia. Revista Brasilira de Ortopedia. 2011;**46**(4):1-8. DOI: 10.1590/S0102-36162011000400003

[18] Domb BG, Dunne KF, Martin TJ, Gui C, Finch NA, Vemula SP, Redmond JM. Patient reported outcomes for patients who returned to sport compared with those who did not after hip arthroscopy: Minimum 2-year follow-up. Journal of Hip Preservation Surgery. 2016;**3**(2):124-131. DOI: 10.1093/jhps/hnv078

[19] Doyon D, Cabanis EA. Diagnóstico por Imagem em Ressonância Magnética. Rio de Janeiro: Medsi; 2000

[20] Erbas M, Simsek T, Kiraz HA, Sahin H, Toman H. Comparação da eficácia de tenoxicam administrado por via oral e intra-articular a pacientes com osteoartrite de joelhos. Revista Brasileira de Anestesiologia. 2015;**65**(5):333-337. DOI: 10.1016/j.bjan.2013.12.003

[21] Feliciano MAR, Canola JC, Vicente WRR. Diagnóstico por imagem em cães e gatos. 1st ed. São Paulo:MedVet; 2015. p. 768

[22] Fernandes GS, Carvalho ACP, Azevedo ACP. Avaliação dos riscos ocupacionais de trabalhadores de serviços de radiologia. Radiologia Brasileira. 2005;**4**(38):279-281. DOI: 10.1590/S0100-39842005000400009

[23] Gattass R, Moll J, Andreiuolo PA, Farias MF, Feitosa PH. Fundamentos da ressonância magnética Funcional. Vol. 13. Cérebro e Mente; 2001 Disponível em:<http://www.epub.org.br/cm

[24] Gonçalves M, Sannomyia EK, Nakazone N, Andréa G. Avaliação de métodos de localização radiográfica para o clínico geral: Parte I. RFO UPF. 2001;**1**(6):45-51

[25] Grant TH, Kelikian AS, Jereb SE. Diagnóstico por ultra-sonografia das rupturas do tendão peroneo. Uma correlação cirúrgica. Journal of Bone and Joint Surgery (American). 2005;**87**(8):1788-1794. DOI: 10.2106/JBJS.D.02450

[26] Grundy EMD. The epidemiology of aging. In: Tallis RC, Fillit HW, editors. Brocklehurst's Textbook of Geriatric Medicine and Gerontology. Philadelphia: Elsevier Science Ltd.; 2003. p. 3-20

[27] Hage MC, Ferrarini NS, Iwasaki M. Imagem por ressonância magnética: princípios básicos. Ciência Rural. 2009;**4**(39):1275-1283. DOI: 10.1590/S0103-84782009005000041

[28] Hochberg M, Lixing L, Bansell B, Langenberg P, Berman B. Traditional Chinese acupuncture is effective as adjunctive therapy in patients with osteoarthritis of the knee. Arthritis Rheumatology. 2004;**50**(1):1-6

[29] Huber M, Trattnig S, Lintner F. Anatomy, biochemistry, and physiology of articular cartilage. Investigative Radiology. 2000;**10**(35):573-580. DOI: 10.1097/00004424-200010000-00003

[30] Hyc A, Osiecka-Iwan A, Jóźwiak J, Moskalewski S. The morphology and selected biological properties of articular cartilage. Ortopedia, Traumatologia, Rehabilitacja. 2001;**2**(3):151-162

[31] Junqueira LC, Carneiro J. Tecido cartilaginoso. In: Junqueira LC, Carneiro J. Histologia básica. 9a ed. Rio de Janeiro: Guanabara Koogan; 2008. 135 p

[32] Khan IM, Willams R, Redman SN, Archer CW. The development of synovial joints. Current Topics in Developmental Biology. 2007;**79**:1-36. DOI: 10.1002/bdrc.10015

[33] Khanna AJ, Cosgarea AJ, Mont MA, Andres BM, Domb BG, Evans PJ, Bluemke DA, Frassica FJ. Magnetic resonance imaging of the knee. Journal of Bone and Joint Surgery. 2001;**83**:128-141 PMID: 11712834

[34] Kidd JA, Fuller C, Barr ARS. Osteoarthritis in the horse. Equine Veterinary Education. 2001;**13**(3):160-168. DOI: 10.1111/j.2042-3292.2001.tb00082.x

[35] Knop PE, Paula LE, Fuller R. Plasma rico em plaquetas no tratamento da osteoartrite. Revista Brasileira de Reumatologia. 2016;**56**(2):152-164

[36] Laredo FJ, Lederman HM, Ihsida A. Doença de Legg-Calvé-Perthes. I-Técnica da artrografia. Revista Brasileira de Ortopedia. 1992;**1**(27):3-6

[37] Loeser RF. The biology of osteoarthritis. In: Annual, Meeting of the American College of Veterinary Pathologistis, Annual Meeting of the American Society for Veterinary Clinical Pathology. Proceedings. v.40: Boston, MA, USA; 2005

[38] Lu P, Kadoya K, Tuszynski MH. Axonal growth and connectivity from neural stem cell grafts in models of spinal cord injury. Current Opinion in Neurobiology. 2014;**1**(27): 103-109

[39] Magee DJ. Avaliação Musculoesquelética. 5a ed. São Paulo:Manole; 2010. p. 1228

[40] Mattos F, Leitea N, Pittab A, Bentoa PCB. Effects of aquatic exercise on muscle strength and functional performance of individuals with osteoarthritis: A systematic review. Revista Brasileira de Reumatologia. 2016;**56**(6):530-542. DOI: 10.1016/j.rbre.2016.09.003 Epub 2016 Oct 4

[41] May SA. Radiological aspects of degenerative joint disease. Equine Veterinary Education. 1999;**8**(2):140-120. DOI: 10.1111/j.2042-3292.1996.tb01861.x

[42] Mazzola AA. Ressonância magnética: princípios de formação da imagem e aplicações em imagem funcional. Revista Brasileira de Física Médica. 2009;**1**(3):117-129

[43] McDonald J, Herzog MM, Philippon MJ. Performance outcomes in professional hockey players following arthroscopic treatment of FAI and microfracture of the hip. Knee Surgery, Sports Traumatology, Arthroscopy. 2014;**22**(4):915-919. DOI: 10.1007/s00167-013-2691-9

[44] Mogami R, JLP V, YFB C, Torezani RS, Vieira AA, ACB K, Barbosa YB, Abreu MM. Ultrasound of ankles in the diagnosis of complications of chikungunya fever. Radiologia Brasileira. 2017;**50**(2):71-75. DOI: 10.1590/0100-3984.2017.50.2e1

[45] Moonen CT, Van Zijl PC, Frank JA, Le Bihan D, Becker ED. Functional magnetic resonance imaging in medicine and physiology. Science. 1990;**250**(4977):53-61

[46] Nagase T, Muneta T, Ju YJ, Hara K, Morito T, Koga H, Nimura A, Mochizuki T, Sekiya I. Analysis of the chondrogenic potential of human synovial stem cells according to harvest site and culture parameters in knees with medial compartment osteoarthritis. Arthritis and Rheumatism. 2008;**58**(5):389-1398. DOI: 10.1002/art.23418

[47] Naredo E, Cabero F, Palop MJ, Collado P, Cruz A, Crespo M. Ultrasonographic findings in knee osteoarthritis: A comparative study with clinical and radiographic assessment. Osteoarthritis and Cartilage. 2005;**13**(7):568-574. DOI: 10.1016/j.joca.2005.02.008

[48] Nobrega AI. Tecnologia radiológica e diagnóstico por Imagem. Editora Difusão: São Paulo; 2006

[49] Ogueta S, Muñoz J, Obregon E, Delgado-baeza E, García-Ruiz JP. Prolactin is a component of the human synovial liquid and modulates the growth and chondrogenic differentiation of bone marrow-derived mesenchymal stem cells. Molecular and Cellular Endocrinology. 2002;**190**(1-2):51-63. DOI: 10.1016/S0303-7207(02)00013-8

[50] Oliveira S. Princípios da artrografia com duplo contraste do joelho. Radiologia Brasileira. 1993;**2**(26):91-97

[51] Paul C, Samdani AF, Betz RR, et al. Grafting of human bone marrow stromal cells into spinal cord injury: A comparison of delivery methods. Spine. 2009;**34**(4):328-334. DOI: 10.1097/BRS.0b013e31819403ce

[52] Pearle AD, Warren RF, Rodeo SA. Basic science of articular cartilage and osteoarthritis. Clinical Sports Medicine. 2005;**24**(1):1-12. DOI: 10.1016/j.csm.2004.08.007

[53] Pelletier JP, Martel-Pelletier J, Howell DS. Etiopathogenesis of osteoarthritis. In: Koopman WJ, editor. Arthritis and Allied Conditions. 14th ed. Philadelphia: Lippincott Williams &Wilkins; 2001. p. 2195-2215

[54] Philippon MJ, Schenker ML. Arthroscopy for the treatment of Femoroacetabular impingement in the athlete. Clinics in Sports Medicine. 2006;**25**(2):299-308. DOI: 10.1016/j.csm.2005.12.006

[55] Rahaman MN, Mao JJ. Stem cell-based composite tissue constructs for regenerative medicine. Biotechnology and Bioengineering. 2005;**91**(3):261-284. DOI: 10.1002/bit.20292

[56] Riggs CM. Osteochondral injury and joint disease in the athletic horse. Equine Veterinary Education. 2006;**18**(2):100-112. DOI: 10.1111/j.2042-3292.2006.tb00426.x.

[57] Rodrigues MB. Diagnostic imaging in musculoskeletal trauma—general principles. Revista de Medicina. 2011;**90**(4):185-194

[58] Schmidt TA, Gastelum NS, Nguyen QT, Schumacher BL, Sah RL. Boundary lubrication of articular cartilage: Role of synovial fluid constituents. Arthritis & Rheumatology. 2007;**56**(3):882-891. DOI: 10.1002/art.22446

[59] Schmitz N, Laverty S, Kraus VB, Aigner T. Basic methods in histopathology of joint tissues. Osteoarthritis and Cartilage. 2010;**1**(18):113-116. DOI: 10.1016/j.joca.2010.05.026.

[60] Siems JJ, Breur GJ, Blevins WE, Cornell KK. Use of two-dimensional realtime ultrasonography for diagnosing contracture and strain of the infraspinatus muscle in a dog. Journal of the American Veterinary Medical Association. 1998;**212**:77-80 PMID: 9426783

[61] Sobotta J. Sobotta Atlas de Anatomia Humana. 23th ed. Rio de Janeiro: Guanabara Koogan; 2012

[62] Sophia FAJ, Bedi A, Rodeo SA. The basic science of articular cartilage: Structure, composition, and function. Sports Health. 2009;**1**(6):461-468. DOI: 10.1177/1941738109350438

[63] Souza ANA, Saladino AO, Biasi C, Matera JM. Uso dos condroprotetores na afecção articular degenerativa: revisão. Revista Acadêmica: Ciências Agrárias e Ambientais. 2010;**3**(8):281-289

[64] Standring S. Osteology. Gray's Anatomy; the Anatomical Basis of Clinical Practice. 40th ed. London: Elsevier Churchill Livingstone; 2010. p. 1433-1439

[65] Thrall DE. Diagnóstico de Radiologia Veterinaria. 6th ed. São Paulo: Elsevier; 2015. p. 848

[66] Tong AC, Tideman H. The microanatomy of the rhesus monkey temporomandibular joint. Journal of Oral and Maxillofacial Surgery. 2001;**59**(1):46-52. DOI: 10.1053/joms.2001.19284

[67] Werner PR, Susko I, Prantoni GA. Regeneração da cartilagem articular lesada experimentalmente em cães em crescimento. Revista do Centro de Ciências Rurais. 2008;**1**(14):59-72

[68] White TD, Black MT, Folkens PA. Human Osteology. 3th ed. Massachusetts:Academic Press; 2011. p. 662

[69] Wight TN. Versican: A versatile extracellular matrix proteoglycan in cell biology. Current Opinion in Cell Biology. 2002;**14**(5):617-623. DOI: 10.1016/S0955-0674(02)00375-7

[70] Witter K, Patulova P, Egerbacher M, Paral V. Morphology of the junction between rib bone and rib cartilage-a discussion of the terms "synchondrosis" and "symphysis". Wiener Tierärztliche Monatsschrift. 2004;**8**(91):214-221

[71] Zhang Z, McCaffery JM, Spencer RGS, Francomano CA. Hyaline cartilage engineered by chondrocytes in pellet culture: Histological, immunohistochemical and ultrastructural analysis in comparison with cartilage explants. Journal of Anatomy. 2004;**205**(3):229-237. DOI: 10.1111/j.0021-8782.2004.00327.x

[72] Zhou S, Cui Z, Urban JP. Factors influencing the oxygen concentration gradient from the synovial surface of articular cartilage to the cartilage-bone interface: A modeling study. Arthritis and Rheumatism. 2004;**50**(12):3915-3924. DOI: 10.1002/art.20675

Chapter 6

Chondral Lesion in the Hip Joint and Current Chondral Repair Techniques

Adrian J. Cassar-Gheiti, Neil G. Burke,
Theresa M. Cassar-Gheiti and Kevin J. Mulhall

Additional information is available at the end of the chapter

http://dx.doi.org/10.5772/intechopen.70261

Abstract

This chapter gives a detailed review of the composition, structure and biomechanics of articular cartilage in the joint. W have looked at the most common types of cartilage lesions and at the existing methods of articular cartilage repair techniques in the hip joint. Articular cartilage is specialized hyaline cartilage which makes a firm, smooth and slippery surface that resists plastic deformation. It has a unique structure and mechanical properties that provide joints with a surface that combines low friction, shock absorption and wear resistance, while bearing large repetitive loads throughout an individual's lifetime. Cartilage lesions in the hip are most common on the acetabular side and typically present as focal area of delamination or chondral flap. Joint preserving techniques are becoming increasingly common. The spectrum of options includes palliative procedures such as joint lavage and chondral debridement, reparative procedures such as microfracture and direct chondral repair, and restorative procedures such as mosaicoplasty. Preservation of the host tissue is most attractive solution to cartilage damage, particularly in young active individuals. Tissue engineering offers one solution but many problems have to be overcome before these techniques become a reality.

Keywords: chondral repair, mosaicoplasty, ACI, MACI, hip joint

1. Introduction

Sports injuries or trauma are a common cause of chondral injuries resulting in joint pain, limitation of function and disability [1]. Articular cartilage is avascular and has very limited capacity for repair [2]. In view of this, chondral lesions that do not penetrate the subchondral bone (partial thickness) do not heal and usually progress to the degeneration of the articular surface [2]. The most common joint affected with chondral injuries is the knee joint [3]. The knee joint accounts for approximately 75% of all reported chondral lesions [4]. In a bibliometric analysis for the most cited topics of arthroscopic procedures, cartilage repair techniques accounted for 53% of all citations, making this the most cited topic in arthroscopic orthopedic surgery and second most cited topic in orthopedics [5]. Cartilage lesions in the hip joint can be due to either traumatic or atraumatic pathologies, these can be associated with labral tears [6, 7], femoroacetabular impingement (FAI) [8], arthritis [9], osteonecrosis and dysplasia [10]. A direct association between acetabular labral injuries and chondral lesions of the femoral head and acetabulum has been reported by various authors [11, 12]. Hip morphology makes chondral injuries in the hip joint difficult to manage, but with recent advances and increased availability of hip arthroscopy over the past years [13], repair techniques commonly applied to the knee joint are being transferred to the hip [14]. Although, in the current literature there is no evidence, early detection and management of chondral lesion may pre-empt degeneration of the entire joint, making hip preserving techniques particularly useful in young active patients.

2. Describing chondral lesion in the hip joint

The hip joint is roughly spherical in shape, but its orientation does not fit exactly. This makes documentation of intra-articular hip lesion challenging. Traditionally a clock face method has been used to topographically report the focus of damage in the hip joint. Although practical the clock face method becomes confusing during arthroscopy and on changing sides. Ilizaliturri et al. [15] have developed and validated an alternative method which is based on anatomical landmarks easily recognizable during arthroscopy (**Figure 1**).

The geographical zone method divides both the acetabulum and the femoral head into six corresponding zones (Zones 1–6) [15]. The acetabulum is divided by two imaginary vertical lines that follow the anterior and posterior limits of the acetabular fossa, which divide it into three sections. A horizontal line perpendicular to the previous lines is placed at the superior limit of the fossa dividing the acetabulum into a superior and inferior part. As a result the acetabulum

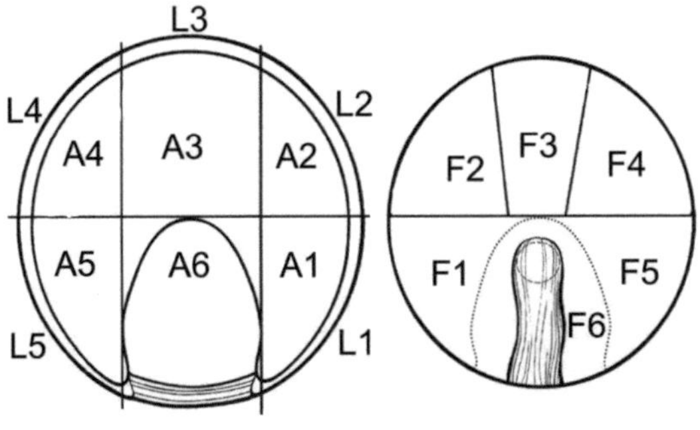

Figure 1. Modified geographical zone mapping system for right acetabulum and right femoral head. Zones A—acetabular zones, Zones L—labral zones, and Zones F—femoral head zones. Adopted from Ilizaliturri et al. [15].

is divided into six zones. Zone 6 corresponds to the fovea on the acetabulum and to the area around the insertion of ligamentum teres on the femoral head. Zone 1 corresponds to the antero-inferior region, Zone 2 to the anterosuperior region, Zone 3 to the central superior region, Zone 4 to the posterosuperior region, Zone 5 to the posteroinferior region on both acetabulum and femoral head while Zone 6 corresponds to the fovea on the acetabulum and the corresponding area around the insertion of ligamentum teres on the femur [15]. This geographical zone method of describing pathology in the hip joint has been used and validated by many authors [6, 16–26].

2.1. Classification for chondral lesions

The spectrum of cartilage damage varies from mild to severe. It is essential to have a reliable classification system for chondral lesion seen during surgery in the hip joint. Most classification for chondral lesions are based on classification used in any other joint [27, 28] but lately new classification are being developed to describe various chondral lesions specific to the hip joint [9, 24]. The most common classification used in the literature is the Outerbridge classification (**Figure 2**) [28] which was described in 1961 and cited 914 times [29] and the second most common classification is the one developed by the International Cartilage Repair Society (ICRS) [27] which was described in 2003 and cited 169 times [29]. **Figure 3** demonstrates the differences between the two classifications.

The Outerbridge classification categorizes chondral injury into four grades from I (slight) to IV (severe) (**Figures 2** and **3**). It is simple and reproducible and new classification systems for the hip joint are based on it [9, 24]. In a Grade I cartilage lesion there is softening or oedema, Grade II there is less than 1.3 cm cartilage fragmentation or tear, Grade III if fragmentation or tear of cartilage is more than 1.3 cm and Grade IV if subchondral bone is visible and breached. New classification systems for the hip joint have taken this further and have described the amount of delamination in the cartilage [9, 24]. For the purpose of this study the Outerbridge classification [28] has been used to describe cartilage injury.

2.2. Type of chondral lesions

Non-arthritic cartilage injuries in the hip refer to focal chondral defects on either the femoral or the acetabular side of the joint. Cartilage lesions in the hip are most common on the

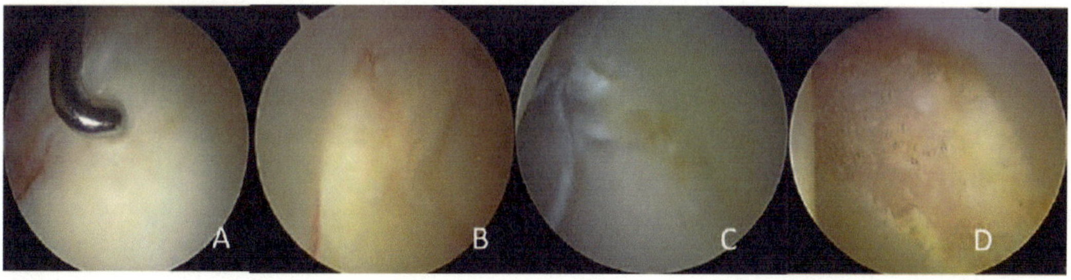

Figure 2. Outerbridge classification during hip arthroscopy. (A)—Grade I, (B)—Grade II, (C)—Grade III, and (D)—Grade IV.

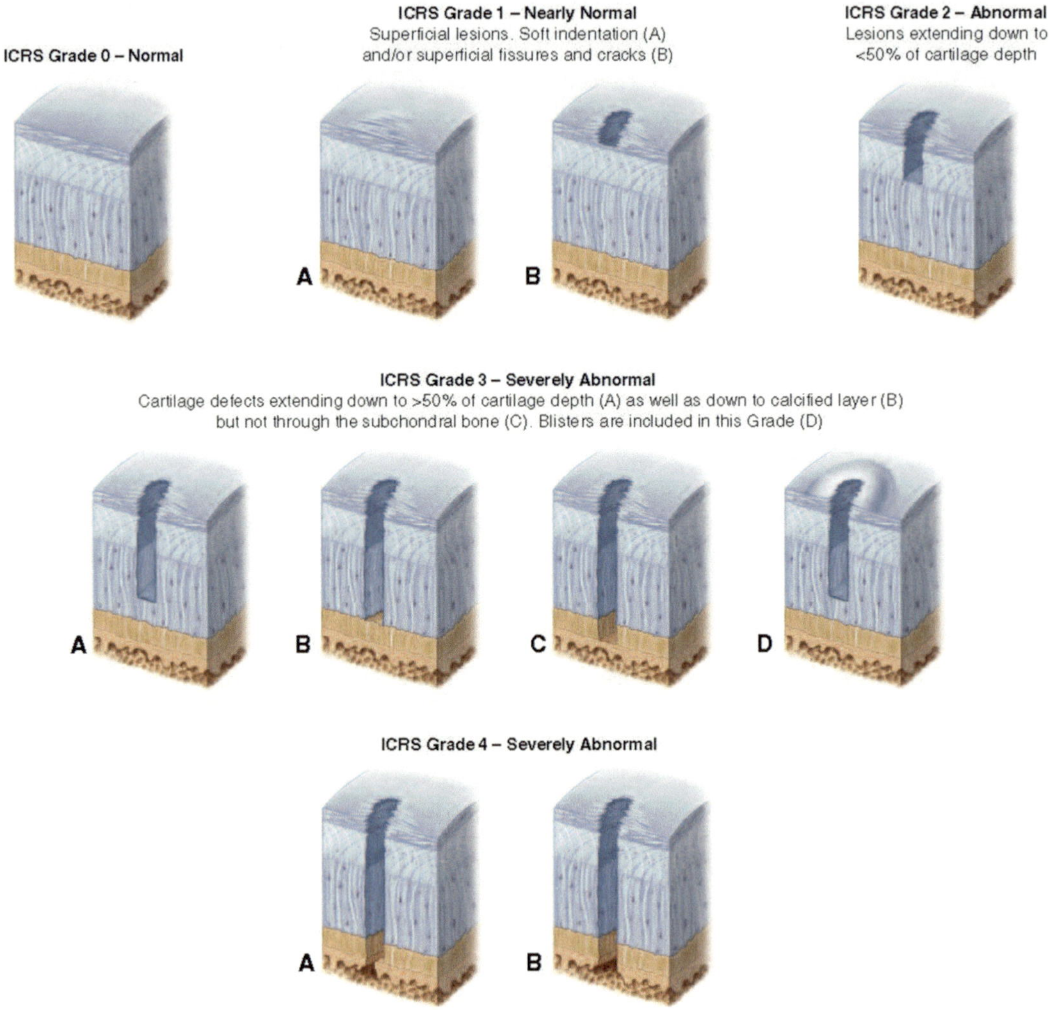

Figure 3. ICRS classification. Adopted from www.cartilage.com.

acetabular side and typically present as focal area of delamination or chondral flap (carpet type lesion). The most common condition resulting in these type of lesions is femoroacetabular impingement (FAI) [30–36]. Most acetabular cartilage lesions are localized to the anterior and anterosuperior region of the acetabulum, present in 59–88% of cases and in the posterior or posterosuperior acetabulum in 25–55% of cases [37]. Lesions on the posterior aspect are commonly related to repetitive posterior loading of the posterior rim of the acetabulum or by axial impact in high energy contact sports [38]. Cartilage lesions on the anterior and anterosuperior aspect are more common in FAI as described by Ganz et al. in both Cam and Pincer type impingement [30–32]. In a series of 273 patients who underwent hip arthroscopy, McCarthy et al. reported that 26% of patient had and Outerbridge IV chondral lesion. They have also reported three distinct patterns Outerbridge IV chondral lesions: (i) isolated lesions with a chondral flap (62%), (ii) localized full-thickness chondral wear without an associated flap (38%), (iii) global degenerative joint disease with areas of full-thickness cartilage loss (6%) (**Figure 4**) [39]. They have also reported that most Grade IV anterior lesions consisted of a chondral flap in continuity with a tear of the articular margin of the labrum. This region was termed the 'watershed zone' by McCarthy et al. [39].

Cartilage lesions on the femoral head are less common, but typically occur from impact loading across the hip joint [33, 40]. Lesions on the femoral head can present as shear injuries, delamination, chondral flaps, fissuring, fractures and impaction injuries. The type and degree of injury depends on the amount and direction of the impact load [33, 38, 40, 41]. Fissuring of cartilage is reported to occur at 25% strain of articular cartilage specimens and the extent of damage to chondrocytes depends on the quality of the underlying bone [42]. In a recent study by Philippon et al. all patients sustained a labral tear and chondral defect following a traumatic hip dislocation. In 14% of the cases an isolated femoral head lesion was observed. Avascular necrosis (AVN) is another known cause of focal cartilage injury to the femoral head, and is secondary to loss of structural integrity of subchondral bone [42]. A wide spectrum of chondral lesions is associated with AVN from mild delamination to complete collapse.

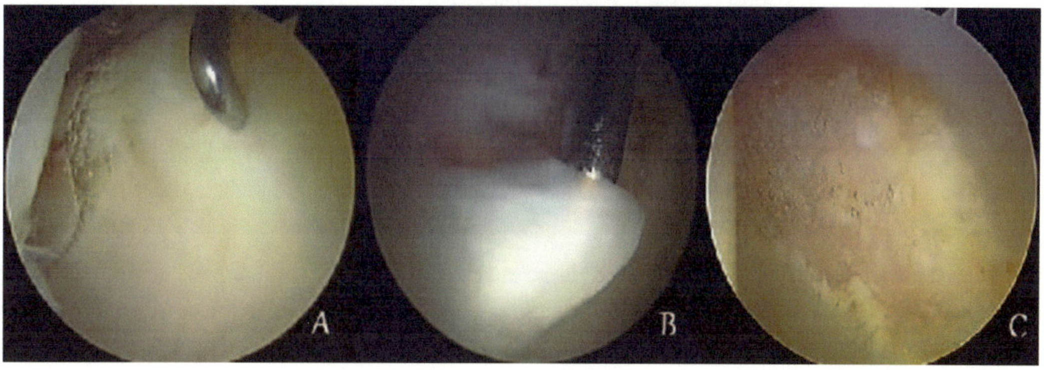

Figure 4. Three different patterns of Grade IV lesions. (A)—Wave sign, (B)—Carpet, and (C)—Global degeneration.

3. Current articular repair techniques

The current goal for surgical intervention is to correct the cause of injury and address the associated chondral pathology. The cause of chondral damage is mostly due to abnormal morphology either the acetabulum or the femoral head and surgery is tailored to the underlying anatomical abnormality. Femoroacetabular impingement is the most common cause of chondral injury in the acetabulum, osteochondroplasty of the femoral neck is one technique used to address this abnormality. Osteochondroplasty only addresses the abnormality on the femoral neck while other techniques are required to repair the associated chondral injury in the acetabulum. Joint-preserving techniques traditionally used in the treatment of cartilage lesions in the knee joint are becoming increasingly utilized in the hip joint. The experience in the hip is limited at this point, but the spectrum of options includes palliative procedures such as joint lavage and chondral debridement, reparative procedures such as microfracture of subchondral bone and recently combined with direct chondral repair [43–47], and restorative procedures such as mosaicoplasty [48], autologous chondrocyte implantation (ACI) [32, 34, 35, 49–59].

3.1. Arthroscopic lavage and debridement

Arthroscopic washout or lavage has been the primary treatment for chondral lesions for the past 24 years [60]. During arthroscopic lavage, inflammatory mediators, loose cartilage and any cartilaginous debris residing in the joint causing synovial inflammation, effusion and bio-mechanical obstruction is washed out. Jackson has reported symptomatic improvement in 45% of patients at 3.5 years and measurable improvement in 80% of patients after arthroscopic lavage [61] with similar results reported by other authors [62, 63]. Most commonly debridement of chondral debris is carried out with arthroscopic lavage. McLaren et al. reported excellent control of pain in 38% of patients and improved function 22% of cases after arthroscopic debridement and lavage [64], similar results were also reported by Gibson et al. [65]. Sözen et al. have reported improvement in Harris Hip Scores (HHS) in 62% of patients after arthroscopic debridement and lavage in osteoarthritis of the hip joint [66]. Arthroscopic lavage and debridement only addresses the patients' symptoms and slow further degeneration by reducing chondral debris in the joint but it does not facilitate defect repair nor does prevent future defect enlargement. Moseley et al. reported no improvement in symptoms or function when arthroscopic lavage and debridement when compared with placebo arthroscopy [67].

3.2. Bone marrow stimulation

Bone marrow stimulation is the most frequent used technique for treating small symptomatic lesions of the articular cartilage in both knee and hip joint. The most common bone marrow stimulation technique is microfracture. This procedure is straightforward and the costs are low compared with other treatment modalities. Microfracture has become increasingly popular among orthopedic surgeons as preferred treatment for chondral defects [45, 50, 68–72].

When subchondral bone is perforated during microfracture it brings undifferentiated stem cells into the defect from the marrow. A marrow clot is established within the microfractured area [68]. The newly formed clot provides an environment for both pluripotent marrow cells and mesenchymal stem cells to differentiate into stable tissue within the base of the lesion [68]. Histological evaluation indicates that fibrocartilaginous tissue is the final product covering the previous lesion [73]. However the overall concentration of mesenchymal stem cells is quite low and declines with age [74]. Reparative fibrocartilage consists of Type-I, Type-II and Type-III in varying amounts and does not resemble the surrounding hyaline cartilage with less Type-II collagen [75, 76].

Phillippon et al. reported that eight of nine patients had 95–100% coverage of an isolated acetabular chondral lesion or acetabular lesion associated with a femoral head lesion, with Grade I or II appearance of the repair product at an average of 20 months follow-up with only one patient progressing to generalized osteoarthritis [55]. Although there are no published long term studies on microfracture in the hip join, studies with good long term results exist for microfracture of the knee [71, 72, 77–79]. Lodhia et al. concluded that microfractures in the hip helps patients to achieve favorable outcomes of their hip with similar results to a matched cohort of patients, who may have a chondral lesion that did not warrant microfracture [46]. Even with meticulous surgical technique and proper patient selection, the results of microfracture appear to deteriorate over time [80]. Although microfracture is an easy reproducible technique that is commonly employed as a first line treatment the results are not as good in older patients and tend to deteriorate over time.

3.3. Direct chondral repair

Direct chondral repair refers to techniques in which a full-thickness chondral flap is repaired back to the subchondral bone rather than debrided. The most recent reported direct chondral repairs are techniques using suture repair [47] and fibrin adhesive [43, 44] in combination with microfracture. These techniques are used on the acetabular side of the hip joint.

3.3.1. Suture repair

This technique describe by Sekiya et al. is used to repair, a chondral flap, where microfractures are applied under the chondral flap. An anchor loaded with absorbable sutures is than fixed in the perilabral sulcus, the suture is passed over the labrum and through the chondral flap, back through the labrum to tie it in the perilabral sulcus [47]. This allows initially stability until the chondral flap heals back in place through fibrosis stimulated by the microfractures. This technique has been only reported by Sekiya et al. and at 2 years follow-up, the patient reported to feel 95% normal, with a Harris Hip Score of 93 and Hip Outcome Score Sports subscale of 81. There are no large studies on this technique available to date and further research is warranted.

3.3.2. Fibrin adhesive repair

Fibrin adhesive is a biological compound, which has been used in many fields of surgery. The haemostatic and adhesive properties of fibrin glue are well known to neurosurgeons [81], ophthalmologists [82, 83], otolaryngologists [84], general [85, 86] and orthopedic surgeons [87, 88]. In orthopedics fibrin adhesive can be used to reattach native hyaline cartilage to the underlying subchondral bone to create an anatomical and durable repair [89]. In the hip joint, Tzaveas et al. reported repair of a chondral flap by using a combination of microfracture and fibrin adhesive under the chondral flap. Follow-up of 43 patients for 1–3 years showed significant improvement in modified Harris Hip Scores with this technique [43]. No randomized control studies of this technique with microfracture or any other technique exists and further studies are required.

3.3.3. Cyanoacrylate

Cyanoacrylates are a class of synthetic glues that rapidly solidify upon contact with weak basis, such as water or blood [90]. Compared with other tissue adhesives cyanoacrylates are easier to use, have quicker polymerization and guarantee higher bonding strength. The use of cyanoacrylate tissue adhesive is well described in the literature for closure of skin wounds [91–93]. Cyanoacrylates is a generic name for a group of tissue adhesives such as ethyl-2-cyanoacrylate, *n*-butyl-2-cyanoacrylate and 2-octyl cyanoacrylate distributed under various names like Histoacryl®, Indermil®, Dermabond® or Glubran®. All cyanoacrylate bond body tissue and show a bacteriostatic effect. In medical practice, *n*-butyl- and octyl-cyanoacrylate are most commonly used. Both biomechanical [94, 95] and cytotoxic [96–98] properties of cyanoacrylate have been tested extensively. *n*-Butyl-2-cyanoacrylate have been approved for internal use including atriovenous embolization [99], endoscopic treatment of bleeding ulcers [100, 101], occlusion of biliary [102] and pancreatic fistulas [103], fixation of polypropylene mesh in open [104, 105] and laparoscopic hernia repair [106]. In orthopedic literature, cyanoacrylate (Dermabond®) has been used for skin closure with excellent result when compared with staples after total joint arthroplasty. A biomechanical study on the use of cyanoacrylate (Histoacryl®) for meniscal repair, reported decrease failure rates when compared to vertical suture repair [95] but no *in vivo* study is yet available. Octyl-cyanoacrylate was used to fix meniscal transplant in a rabbit model, the authors had to sacrifice all animals earlier than planned due to severe inflammatory reaction with caseous necrosis in the operated joint and they have recommended against the use of octyl-cyanoacrylate to fix transplanted menisci [107]. A new cyanoacrylate, 'Glubran 2' (GEM Srl, Viareggio, Italy) is authorized for surgical use and with a CE mark for 'internal use'. Glubran 2 is different to other cyanoacrylates as it has a different chemical composition making it a co-monomer rather than a simple monomer and is composed of *n*-butyl-2-cyanoacrylate and methacryloxysulfolane monomer [104]. The difference in compositions, allows polymerization at lower temperatures and reduced inflammatory reaction when compared to other cyanoacrylates [97, 108]. In recent years a number of clinical studies in general surgery have reported good results when 'Glubran 2' has been used *in vivo* [104–106]. At this stage there is no clinical study evaluating the use of cyanoacrylate intraarticularly.

Biomechanical data published on chondral repair techniques has shown improve resistance to shear forces across the chondral surface when compared to fibrin adhesive repair in cadaveric models [109]. Furthermore we have identified early biomechanical failure in fibrin adhesive repair, which failed at only 50 cycles, while suture of chondral flaps were more biomechanically stable throughout the 1500 cycle testing [109]. The small number of reported outcomes and early laboratory failure may limit fibrin glue clinical use, however, both fibrin glue, suture and cyanoacrylate repair warrant further investigation.

3.4. Whole tissue transplantation

The use of whole tissue chondral transplantation using either an autograft or an allograft is well known in the orthopedics [56, 110–115].

In autologous osteochondral transplantation, occasionally referred as osteoarticular transfer system (OATS), is an effective method for resurfacing osteochondral defects and most commonly used in the knee joint. This technique involves transplantation of multiple cylindrical osteochondral plugs harvested from a non-weight or less weight bearing areas of the articular surface in the joint and transferred to create a congruent and durable area in the defect. Koh et al. assessed contact pressures on a swine knee model and reported that flush or slightly sunk grafts could restore contact pressures to nearly normal levels, but elevated angled grafts adversely increased contact pressures [116]. However, they used only one plug, which does not correlate with clinical practice. Kock et al. reported reduction in contact pressures after OATS to be 30% less than contact pressures before the procedure with an empty defect in a human cadaveric knee [117]. The outcomes of autologous mosaicoplasty are promising, Hangody and Füles evaluated the largest series of mosaicoplasty performed for localized Outerbridge Grade III or IV lesions and reported good to excellent results for 92% of the femoral lesions, 87% of tibial lesions and 79% of patellofemoral lesions [118]. Ollat et al. reported satisfactory results in 72.5% of the patients at 8 years of follow-up and that the largest defects with the longest follow-up have the worst prognosis [111]. Osteochondral mosaicoplasty of the femoral head has mixed prognosis; Rittmeister et al. reported that four out of five hips had unsatisfactory results after 5 years follow-up and underwent total hip arthroplasty [119], while Girard et al. reported satisfactory improvements in Postel Merle d'Aubingé Score and global range of motion in the hip joint at an average follow-up of 30 months [120]. Nam et al. reported on two cases that underwent OATS combined with osteochondral fragment fixation after traumatic anterior dislocation of the hip joints [121]. They showed good clinical outcomes and graft incorporation using magnetic resonance imaging (MRI) [121]. Emre et al. have reported good, pain free results a 3 years after surgery [122]. Good clinical outcomes were also reported for fragment fixation combined with OATS for the treatment of osteochondral defects after posterior fracture-dislocation of the hip joint [123]. Recently, good results have been reported from arthroscopic OATS procedure in one patients with 2 year follow-up [48, 124]. Arthroscopic OATS procedures for treating osteochondral lesions of the femoral head are promising but more studies with more patients and longer follow-up periods are required to fully understand the benefits of mosaicoplasty in the hip joint.

Osteochondral allograft transplantation is chondral surface reconstruction that involves transplantation of a cadaveric graft consisting of intact, viable articular cartilage and its underlying subchondral bone into the defect. Currently fresh osteochondral allografts are utilized to treat a broad spectrum of articular cartilage pathology, from focal chondral defects to joints with established osteoarthritis in the hip, knee and ankle joint [125–127]. Advantages to the use of osteochondral allografts include the ability to achieve precise surface architecture, immediate transplantation of viable hyaline cartilage, the potential to replace large defects and no donor site morbidity. Like any allograft transplantation, limitations include; limited graft availability, high cost, risk of immunological reactions and rejections, potential for disease transmission and technically demanding aspect of machining and sizing the allograft [128]. A number of retrospective studies have been performed to assess the outcomes of osteochondral allograft transplantation for the treatment of focal osteochondral defects of the knee, and they have demonstrated good-to-excellent results [129–132]. Krych et al. have reported improvement in Harris Hip Score at 2 and 3 year follow-up in two cases that underwent osteochondral allograft of the acetabulum [113]. Gross et al. reported survival rates for osteochondral allografts of 95% at five years, 85% at 10 years and 73% at fifteen years for posttraumatic femoral condylar lesions [133].

3.5. Cell based and scaffold treatment

Autologous chondrocyte implantation (ACI) was originally described by Brittberg et al. [134]. ACI is an innovative technique to restore cartilage cells into full-thickness chondral defects. In ACI there is development of hyaline like cartilage rather than fibrocartilage in the defect, leading to better long term outcomes and longevity of the healing tissue. Good out comes have been reported by various authors. ACI involves two surgical procedures, the first operation is used to harvest the tissue required and the second procedure is required to implant the chondrocytes in the defect. During the second procedure periosteal is also harvested from a different site and used to contain the chondrocytes in the chondral defect. ACI is not without limitations; not many patients are willing to undergo two procedures and there is a risk of donor site morbidity at the periosteal harvest site. Adverse events after ACI have been reported in 46% of patients undergoing the procedure, with graft failure accounting for 25%, delamination accounting for 22% and tissue hypertrophy occurred in about 18% of cases [135]. Peterson et al. reported 52 adverse events, including 26 instances of periosteal hypertrophy and seven graft failure in 101 patients [136].

In second generation or scaffold based ACI, harvested chondrocytes are delivered on an absorbable scaffold that supports the cells preimplantation culturing and postoperative healing process. In matrix-associated chondrocyte implantation (MACI) procedure chondrocytes are incorporated into various types of tissue engineered scaffolds. Various tissue-engineered compounds are being used as scaffolds including hyaluronan, alginates, agarose hydrogels and gelatin scaffolds [137–140]. The results from MACI to treat chondral defects have been encouraging, Behrens et al. reported substantial improvement in clinical outcome scores in 35% of patients at 5 year follow-up [141]. Marcacci et al. reported improvement in quality of life as assessed by the EuroQol - Visual Analogue Scale (EQ-VAS) in 93% of patient at 2 year follow-up after hyaluronan-based scaffold MACI, with resumption of sports at same or slightly lower level in 56.7% of patients at 12 months [142]. Although promising results are being reported after MACI, long term clinical outcomes associated with this procedure are still limited.

The autologous matrix-induced chondrogenesis (AMIC), further develops the scaffold technique in combination with micro-fracturing [59]. It is a one-step procedure that involves microfracturing of the debrided cartilage lesion and a commercially available collagen I/III matrix for covering the blood clot and its MSCs. Fixation is with partial autologous fibrin glue in which the thrombin part is yielded from the patient's serum. The indications of AMIC are symptomatic full-thickness chondral and subchondral defects in the major joints, maximum size of 2–4 cm^2, posttraumatic or osteochondrosis dissecans, and location in the main weight bearing area of the joint or maximum area of pain [59, 143]. In one study, patients with large Grade IV chondral lesions experienced significant improvement up to 24 months after the AMIC procedure [144]. Recently, Fontana has reported on the 5 year follow-up of 201 patients treated with AMIC in the hip joint. This study reported continuous improvement with respect to each evaluation time point in modified Harris Hip Scores peaking at 3 years follow-up [59]. The AMIC technique is further beneficial because it eliminates the need for specialized centers and laboratory support to cultivate cells, in turn reducing total therapy time and overall cost, compared to twostage procedures such as MACI.

4. Conclusion

Management of chondral lesion the hip joint to preserve the native joint in young active patients with chondral lesion is challenging for the orthopedic surgeon. Joint-preserving technique in the hip joint continue to evolve with recent reports showing promising results. Indications for these techniques continue to expand and a simplified algorithm was proposed by El Bitar et al. for join preserving management of articular cartilage lesions in the hip joint [14]. The literature so far is limited to low evidence studies with lack of control groups making comparison of different treatment options difficult. Further research in these different modalities is required to formulate a best treatment practice guidelines in the treatment of chondral lesions in the hip.

Author details

Adrian J. Cassar-Gheiti*, Neil G. Burke, Theresa M. Cassar-Gheiti and Kevin J. Mulhall

*Address all correspondence to: adriancassargheiti@gmail.com

Cappagh National Orthopaedic Hospital, Dublin, Ireland

References

[1] Peters CL, Erickson J. The etiology and treatment of hip pain in the young adult. Journal of Bone & Joint Surgery – American Volume. 2006;**88**(Suppl 4):20-26

[2] Rolauffs B, et al. Vulnerability of the superficial zone of immature articular cartilage to compressive injury. Arthritis & Rheumatology. 2010;**62**(10):3016-3027

[3] Flanigan DC, et al. Prevalence of chondral defects in athletes' knees: A systematic review. Medicine & Science in Sports & Exercise. 2010;**42**(10):1795-1801

[4] Obedian RS, Grelsamer RP. Osteochondritis dissecans of the distal femur and patella. Clinics in Sports Medicine. 1997;**16**(1):157-174

[5] Cassar Gheiti AJ, et al. The 25 most cited articles in arthroscopic orthopaedic surgery. Arthroscopy. 2012;**28**(4):548-564

[6] Sampson TG. Arthroscopic treatment for chondral lesions of the hip. Clinics in Sports Medicine. 2011;**30**(2):331-348

[7] Guanche CA, Sikka RS. Acetabular labral tears with underlying chondromalacia: A possible association with high-level running. Arthroscopy. 2005;**21**(5):580-585

[8] Bare AA, Guanche CA. Hip impingement: The role of arthroscopy. Orthopedics. 2005;**28**(3):266-273

[9] Beck M, et al. Hip morphology influences the pattern of damage to the acetabular cartilage: Femoroacetabular impingement as a cause of early osteoarthritis of the hip. Journal of Bone & Joint Surgery – British Volume. 2005;**87**(7):1012-1018

[10] Reijman M, et al. Acetabular dysplasia predicts incident osteoarthritis of the hip: The Rotterdam study. Arthritis & Rheumatology. 2005;**52**(3):787-793

[11] Byrd JW. Labral lesions: An elusive source of hip pain case reports and literature review. Arthroscopy. 1996;**12**(5):603-612

[12] McCarthy JC, et al. The Otto E. Aufranc Award: The role of labral lesions to development of early degenerative hip disease. Clinical Orthopaedics and Related Research. 2001;**393**:25-37

[13] Colvin AC, Harrast J, Harner C. Trends in hip arthroscopy. Journal of Bone & Joint Surgery – American Volume. 2012;**94**(4):e23

[14] El Bitar YF, et al. Joint-preserving surgical options for management of chondral injuries of the hip. Journal of the American Academy of Orthopaedic Surgeons. 2014;**22**(1):46-56

[15] Ilizaliturri Jr VM, et al. A geographic zone method to describe intra-articular pathology in hip arthroscopy: Cadaveric study and preliminary report. Arthroscopy. 2008;**24**(5):534-539

[16] Shindle MK, et al. Arthroscopic management of labral tears in the hip. Journal of Bone & Joint Surgery – American Volume, 2008;**90**(Suppl 4):2-19

[17] Larson CM, Giveans MR. Arthroscopic debridement versus refixation of the acetabular labrum associated with femoroacetabular impingement. Arthroscopy. 2009;**25**(4):369-376

[18] Sampatchalit S, et al. Changes in the acetabular fossa of the hip: MR arthrographic findings correlated with anatomic and histologic analysis using cadaveric specimens. American Journal of Roentgenology. 2009;**193**(2):W127-W133

[19] Ruiz-Suarez M, Aziz-Jacobo J, Barber FA. Cyclic load testing and ultimate failure strength of suture anchors in the acetabular rim. Arthroscopy. 2010;**26**(6):762-768

[20] Blankenbaker DG, et al. MR arthrography of the hip: Comparison of IDEAL-SPGR volume sequence to standard MR sequences in the detection and grading of cartilage lesions. Radiology. 2011;**261**(3):863-871

[21] Colvin AC, Koehler SM, Bird J. Can the change in center-edge angle during pincer trimming be reliably predicted? Clinical Orthopaedics and Related Research. 2011;**469**(4):1071-1074

[22] Cross MB, et al. Impingement (acetabular side). Clinics in Sports Medicine. 2011;**30**(2):379-390

[23] Ilizaliturri Jr VM, et al. Hip arthroscopy after traumatic hip dislocation. The American Journal of Sports Medicine. 2011;**39**(Suppl):50S-57S

[24] Konan S, et al. Validation of the classification system for acetabular chondral lesions identified at arthroscopy in patients with femoroacetabular impingement. Journal of Bone & Joint Surgery – British Volume. 2011;**93**(3):332-336

[25] Sendtner E, Winkler R, Grifka J. Femoroacetabular impingement: Minimally invasive hip surgery. Orthopade. 2011;**40**(3):261-270; quiz 271

[26] Gerhardt M, et al. Characterisation and classification of the neural anatomy in the human hip joint. HIP International. 2012;**22**(1):75-81

[27] Brittberg M, Winalski CS. Evaluation of cartilage injuries and repair. Journal of Bone & Joint Surgery – American Volume. 2003;**85-A**(Suppl 2):58-69

[28] Outerbridge RE. The etiology of chondromalacia patellae. Journal of Bone & Joint Surgery – British Volume. 1961;**43-B**:752-757

[29] ISI Web Of Knowledge. 2012, Thomas Reuters. www.webofknowledge.com

[30] Beck M, et al. Anterior femoroacetabular impingement: Part II. Midterm results of surgical treatment. Clinical Orthopaedics and Related Research. 2004;**418**:67-73

[31] Ganz R, et al. Femoroacetabular impingement: A cause for osteoarthritis of the hip. Clinical Orthopaedics and Related Research. 2003;**417**:112-120

[32] Lavigne M, et al. Anterior femoroacetabular impingement: Part I. Techniques of joint preserving surgery. Clinical Orthopaedics and Related Research. 2004;**418**:6166

[33] Philippon MJ, et al. Arthroscopic findings following traumatic hip dislocation in 14 professional athletes. Arthroscopy. 2009;**25**(2):169-174

[34] Clohisy JC, et al. AOA symposium. Hip disease in the young adult: Current concepts of etiology and surgical treatment. Journal of Bone & Joint Surgery – American Volume. 2008;**90**(10):2267-2281

[35] Shindle MK, et al. Hip arthroscopy in the athletic patient: Current techniques and spectrum of disease. Journal of Bone & Joint Surgery – American Volume. 2007;**89**(Suppl 3):29-43

[36] Singh PJ, O'Donnell JM. The outcome of hip arthroscopy in Australian football league players: A review of 27 hips. Arthroscopy. 2010;**26**(6):743-749

[37] Schmid MR, et al. Cartilage lesions in the hip: Diagnostic effectiveness of MR arthrography. Radiology. 2003;**226**(2):382-386

[38] Moorman 3rd CT, et al. Traumatic posterior hip subluxation in American football. Journal of Bone & Joint Surgery – American Volume. 2003;**85-A**(7):1190-1196

[39] McCarthy JC, et al. The watershed labral lesion: Its relationship to early arthritis of the hip. Journal of Arthroplasty. 2001;**16**(8 Suppl 1):81-87

[40] Byrd JW. Lateral impact injury. A source of occult hip pathology. Clinics in Sports Medicine. 2001;**20**(4):801-815

[41] Schmitt KU, Schlittler M, Boesiger P. Biomechanical loading of the hip during side jumps by soccer goalkeepers. Journal of Sports Sciences. 2010;**28**(1):53-59

[42] Krueger JA, et al. The extent and distribution of cell death and matrix damage in impacted chondral explants varies with the presence of underlying bone. Journal of Biomechanical Engineering. 2003;**125**(1):114-119

[43] Stafford GH, Bunn JR, Villar RN. Arthroscopic repair of delaminated acetabular articular cartilage using fibrin adhesive. Results at one to three years. HIP International. 2011;**21**(6):744-750

[44] Tsaveas AP, Villar RN, Arthroscopic repair of acetabular chondral delamination with fibrin adhesive. HIP International. 2010;**20**(1):115-119

[45] McGill KC, Bush-Joseph CA, Nho SJ. Hip microfracture: Indications, technique, and outcomes. Cartilage. 2010;**1**(2):127-136

[46] Lodhia P, et al. Microfracture in the hip: A matched-control study with average 3-year follow-up. Journal of Hip Preservation Surgery. 2015;**2**(4):417-427

[47] Sekiya JK, Martin RL, Lesniak BP. Arthroscopic repair of delaminated acetabular articular cartilage in femoroacetabular impingement. Orthopedics. 2009;**32**(9). DOI: 10.3928/01477447-20090728-44

[48] Kubo T, et al. Hip arthroscopic osteochondral autologous transplantation for treating osteochondritis dissecans of the femoral head. Arthroscopy Techniques. 2015;**4**(6): e675-e680

[49] Akimau P, et al. Autologous chondrocyte implantation with bone grafting for osteochondral defect due to posttraumatic osteonecrosis of the hip – A case report. Acta Orthopaedica. 2006;**77**(2):333-336

[50] Crawford K, et al. Microfracture of the hip in athletes. Clinics in Sports Medicine. 2006;**25**(2):327-335, x

[51] Hart R, et al. Mosaicplasty for the treatment of femoral head defect after incorrect resorbable screw insertion. Arthroscopy. 2003;**19**(10):E1-E5

[52] Millis MB, Kim YJ. Rationale of osteotomy and related procedures for hip preservation: A review. Clinical Orthopaedics and Related Research. 2002;**405**:108-121

[53] Nousiainen MT, et al. The use osteochondral allograft in the treatment of a severe femoral head fracture. Journal of Orthopaedic Trauma. 2010;**24**(2):120-124

[54] Parvizi J, et al. Management of arthritis of the hip in the young adult. Journal of Bone & Joint Surgery – British Volume. 2006;**88**(10):1279-1285

[55] Philippon MJ, et al. Can microfracture produce repair tissue in acetabular chondral defects? Arthroscopy. 2008;**24**(1):46-50

[56] Williams RJ, editor. Cartilage Repair Strategies. Totowa, NJ: Humana Press; 2007. xvii, 374 p

[57] Fontana A, et al. Arthroscopic treatment of hip chondral defects: Autologous chondrocyte transplantation versus simple debridement – A pilot study. Arthroscopy. 2012;**28**(3):322-329

[58] Knutsen G, et al. A randomized multicenter trial comparing autologous chondrocyte implantation with microfracture: Long-term follow-up at 14 to 15 years. Journal of Bone & Joint Surgery – American Volume. 2016;**98**(16):1332-1339

[59] Fontana A. Autologous Membrane Induced Chondrogenesis (AMIC) for the treatment of acetabular chondral defect. Muscles, Ligaments and Tendons Journal. 2016;**6**(3): 367-371

[60] Bauer M, Jackson RW. Chondral lesions of the femoral condyles: A system of arthroscopic classification. Arthroscopy. 1988;**4**(2):97-102

[61] Jackson RW. Arthroscopic Treatment of degenerative Arthritis, In: McGinty JB, editor. Operative Arhroscopy. New York: Raven press; 1991

[62] Chang RW, et al. A randomized, controlled trial of arthroscopic surgery versus closed-needle joint lavage for patients with osteoarthritis of the knee. Arthritis & Rheumatology. 1993;**36**(3):289-296

[63] Livesley PJ, et al. Arthroscopic lavage of osteoarthritic knees. Journal of Bone & Joint Surgery – British Volume. 1991;**73**(6):922-926

[64] McLaren AC, et al. Arthroscopic debridement of the knee for osteoarthrosis. Canadian Journal of Surgery. 1991;**34**(6):595-598

[65] Gibson JN, et al. Arthroscopic lavage and debridement for osteoarthritis of the knee. Journal of Bone & Joint Surgery – British Volume. 1992;**74**(4):534-537

[66] Sozen YV, et al. The effectiveness of arthroscopic debridement and lavage treatment in osteoarthritis of the hip: Preliminary results. Acta Orthopaedica et Traumatologica Turcica. 2004;**38**(2):96-103

[67] Moseley JB, et al. A controlled trial of arthroscopic surgery for osteoarthritis of the knee. New England Journal of Medicine. 2002;**347**(2):81-88

[68] Steadman JR, Rodkey WG, Rodrigo JJ. Microfracture: Surgical technique and rehabilitation to treat chondral defects. Clinical Orthopaedics and Related Research. 2001;**391**(Suppl): S362-S369

[69] Steadman JR, et al. The microfracture technic in the management of complete cartilage defects in the knee joint. Orthopade. 1999;**28**(1):26-32

[70] Steadman JR, Rodkey WG, Briggs KK. Microfracture: Its history and experience of the developing surgeon. Cartilage. 2010;**1**(2):78-86

[71] Steadman JR, et al. Outcomes of microfracture for traumatic chondral defects of the knee: Average 11-year follow-up. Arthroscopy. 2003;**19**(5):477-484

[72] Knutsen G, et al. Autologous chondrocyte implantation compared with microfracture in the knee. A randomized trial. Journal of Bone & Joint Surgery – American Volume. 2004;**86-A**(3):455-464

[73] Frisbie DD, et al. Early events in cartilage repair after subchondral bone microfracture. Clinical Orthopaedics and Related Research. 2003;**407**:215-227

[74] Tran-Khanh N, et al. Aged bovine chondrocytes display a diminished capacity to produce a collagen-rich, mechanically functional cartilage extracellular matrix. Journal of Orthopaedic Research. 2005;**23**(6):1354-1362

[75] Frisbie DD, et al. Arthroscopic subchondral bone plate microfracture technique augments healing of large chondral defects in the radial carpal bone and medial femoral condyle of horses. Veterinary Surgery. 1999;**28**(4):242-255

[76] Bae DK, Yoon KH, Song SJ. Cartilage healing after microfracture in osteoarthritic knees. Arthroscopy. 2006;**22**(4):367-374

[77] Saris DB, et al. Characterized chondrocyte implantation results in better structural repair when treating symptomatic cartilage defects of the knee in a randomized controlled trial versus microfracture. The American Journal of Sports Medicine. 2008;**36**(2):235-246

[78] Gudas R, et al. A prospective randomized clinical study of mosaic osteochondral autologous transplantation versus microfracture for the treatment of osteochondral defects in the knee joint in young athletes. Arthroscopy. 2005;**21**(9):1066-1075

[79] Gobbi A, Nunag P, Malinowski K. Treatment of full thickness chondral lesions of the knee with microfracture in a group of athletes. Knee Surgery, Sports Traumatology, Arthroscopy. 2005;**13**(3):213-221

[80] Mithoefer K, et al. High-impact athletics after knee articular cartilage repair: A prospective evaluation of the microfracture technique. The American Journal of Sports Medicine. 2006;**34**(9):1413-1418

[81] Jankowitz BT, et al. Effect of fibrin glue on the prevention of persistent cerebral spinal fluid leakage after incidental durotomy during lumbar spinal surgery. European Spine Journal. 2009;**18**(8):1169-1174

[82] Lagoutte FM, Gauthier L, Comte PR. A fibrin sealant for perforated and preperforated corneal ulcers. British Journal of Ophthalmology. 1989;**73**(9):757-761

[83] Shehadeh-Mashor R, et al. Management of recurrent pterygium with intraoperative mitomycin C and conjunctival autograft with fibrin glue. American Journal of Ophthalmology. 2011;**152**(5):730-732

[84] Hobbs CG, Darr A, Carlin WV. Management of intra-operative cerebrospinal fluid leak following endoscopic trans-sphenoidal pituitary surgery. Journal of Laryngology & Otology. 2011;**125**(3):311-313

[85] Campanelli G, et al. Randomized, controlled, blinded trial of Tisseel/Tissucol for mesh fixation in patients undergoing Lichtenstein technique for primary inguinal hernia repair: Results of the TIMELI trial. Annals of Surgery. 2012;**255**(4):650-657

[86] Fortelny RH, et al. Use of fibrin sealant (Tisseel/Tissucol) in hernia repair: A systematic review. Surgical Endoscopy. 2012;**26**(7):1803-1812

[87] Massin P, et al. Does fibrin sealant use in total knee replacement reduce transfusion rates? A non-randomised comparative study. Orthopaedics & Traumatology: Surgery & Research. 2012;**98**(2):180-185

[88] Bekkers JE, et al. Quality of scaffold fixation in a human cadaver knee model. Osteoarthritis and Cartilage. 2010;**18**(2):266-272

[89] Shah MA, Ebert AM, Sanders WE. Fibrin glue fixation of a digital osteochondral fracture: Case report and review of the literature. Journal of Hand Surgery American Society. 2002;**27**(3):464-469

[90] Esposito C, et al. Experience with the use of tissue adhesives in pediatric endoscopic surgery. Surgical Endoscopy. 2004;**18**(2):290-292

[91] Quinn J, et al. A randomized trial comparing octylcyanoacrylate tissue adhesive and sutures in the management of lacerations. Journal of the American Medical Association. 1997;**277**(19):1527-1530

[92] Qureshi A, et al. n-Butyl cyanoacrylate adhesive for skin closure of abdominal wounds: Preliminary results. Annals of the Royal College of Surgeons of England. 1997;**79**(6):414-415

[93] Liebelt EL. Current concepts in laceration repair. Current Opinion in Pediatrics. 1997;**9**(5): 459-464

[94] Kull S, et al. Glubran 2 surgical glue: In vitro evaluation of adhesive and mechanical properties. Journal of Surgical Research. 2009;**157**(1):e15-e21

[95] Ayan I, et al. Histoacryl glue in meniscal repairs (a biomechanical study). International Orthopaedics. 2007;**31**(2):241-246

[96] Papatheofanis FJ. Cytotoxicity of alkyl-2-cyanoacrylate adhesives. Journal of Biomedical Materials Research. 1989;**23**(6):661-668

[97] Montanaro L, et al. Cytotoxicity, blood compatibility and antimicrobial activity of two cyanoacrylate glues for surgical use. Biomaterials. 2001;**22**(1):59-66

[98] Evans CE, Lees GC, Trail IA. Cytotoxicity of cyanoacrylate adhesives to cultured tendon cells. Journal of Hand Surgery: British & European Volume. 1999;**24**(6):658-661

[99] n-BCA Trial Investigtors. N-butyl cyanoacrylate embolization of cerebral arteriovenous malformations: Results of a prospective, randomized, multi-center trial. American Journal of Neuroradiology. 2002;**23**(5):748-755

[100] Dhiman RK, et al. Endoscopic sclerotherapy of gastric variceal bleeding with N-butyl-2-cyanoacrylate. Journal of Clinical Gastroenterology. 2002;**35**(3):222-227

[101] Seewald S, et al. Cyanoacrylate glue in gastric variceal bleeding. Endoscopy. 2002;**34**(11): 926-932

[102] Seewald S, et al. Endoscopic treatment of biliary leakage with n-butyl-2 cyanoacrylate. Gastrointestinal Endoscopy. 2002;**56**(6):916-919

[103] Mutignani M, et al. External pancreatic fistulas resistant to conventional endoscopic therapy: Endoscopic closure with N-butyl-2-cyanoacrylate (Glubran 2). Endoscopy. 2004;**36**(8):738-742

[104] Testini M, et al. A single-surgeon randomized trial comparing sutures, N-butyl-2-cyanoacrylate and human fibrin glue for mesh fixation during primary inguinal hernia repair. Canadian Journal of Surgery. 2010;**53**(3):155-160

[105] Paajanen H, et al. Randomized clinical trial of tissue glue versus absorbable sutures for mesh fixation in local anaesthetic Lichtenstein hernia repair. British Journal of Surgery. 2011;**98**(9):1245-1251

[106] Kukleta JF, Freytag C, Weber M. Efficiency and safety of mesh fixation in laparoscopic inguinal hernia repair using n-butyl cyanoacrylate: Long-term biocompatibility in over 1300 mesh fixations. Hernia. 2012;**16**(2):153-162

[107] Reckers LJ, Fagundes DJ, Cohen M. The ineffectiveness of fibrin glue and cyanoacrylate on fixation of meniscus transplants in rabbits. Knee. 2009;**16**(4):290-294

[108] Levrier O, et al. Efficacy and low vascular toxicity of embolization with radical versus anionic polymerization of n-butyl-2-cyanoacrylate (NBCA). An experimental study in the swine. Journal of Neuroradiology. 2003;**30**(2):95-102

[109] Cassar-Gheiti AJ, et al. Comparison of four chondral repair techniques in the hip joint: A biomechanical study using a physiological human cadaveric model. Osteoarthritis and Cartilage. 2015;**23**(6):1018-1025

[110] Krusche-Mandl I, et al. Long-term results 8 years after autologous osteochondral transplantation: 7 T gagCEST and sodium magnetic resonance imaging with morphological and clinical correlation. Osteoarthritis and Cartilage. 2012;**20**(5):357-363

[111] Ollat D, et al. Mosaic osteochondral transplantations in the knee joint, midterm results of the SFA multicenter study. Orthopaedics & Traumatology: Surgery & Research. 2011; **97**(8 Suppl):S160-S166

[112] Robert H. Chondral repair of the knee joint using mosaicplasty. Orthopaedics & Traumatology: Surgery & Research. 2011;**97**(4):418-429

[113] Krych AJ, Lorich DG, Kelly BT. Treatment of focal osteochondral defects of the acetabulum with osteochondral allograft transplantation. Orthopedics. 2011;**34**(7):e307-e311

[114] Scully WF, Parada SA, Arrington ED. Allograft osteochondral transplantation in the knee in the active duty population. Military Medicine. 2011;**176**(10):1196-1201

[115] Krych AJ, et al. Return to athletic activity after osteochondral allograft transplantation in the knee. The American Journal of Sports Medicine. Am J Sports Med. 2012 May;**40**(5):1053-9

[116] Koh JL, et al. The effect of graft height mismatch on contact pressure following osteochondral grafting: A biomechanical study. The American Journal of Sports Medicine. 2004;**32**(2):317-320

[117] Kock NB, et al. A cadaveric analysis of contact stress restoration after osteochondral transplantation of a cylindrical cartilage defect. Knee Surgery, Sports Traumatology, Arthroscopy. 2008;**16**(5):461-468

[118] Hangody L, Füles P. Autologous osteochondral mosaicplasty for the treatment of full-thickness defects of weight-bearing joints: Ten years of experimental and clinical experience. Journal of Bone & Joint Surgery – American Volume. 2003;**85-A**(Suppl 2): 25-32

[119] Rittmeister M, et al. Five-year results following autogenous osteochondral transplantation to the femoral head. Orthopade. 2005;**34**(4):320, 322-326

[120] Girard J, et al. Osteochondral mosaicplasty of the femoral head. HIP International. 2011;**21**(5):542-548

[121] Nam D, et al. Traumatic osteochondral injury of the femoral head treated by mosaicplasty: A report of two cases. The Musculoskeletal Journal of Hospital for Special Surgery. 2010;**6**(2):228-234

[122] Emre TY, et al. Mosaicplasty for the treatment of the osteochondral lesion in the femoral head. Bulletin of the NYU Hospital for Joint Diseases. 2012;**70**(4):288-290

[123] Gagala J, Tarczynska M, Gaweda K, Fixation of femoral head fractures with autologous osteochondral transfer (mosaicplasty). Journal of Orthopaedic Trauma. 2014;**28**(9):e226-e230

[124] Cetinkaya S, Toker B, Taser O. Arthroscopic retrograde osteochondral autologous transplantation to chondral lesion in femoral head. Orthopedics. 2014;**37**(6): e600-e604

[125] Evans KN, Providence BC. Case report: Fresh-stored osteochondral allograft for treatment of osteochondritis dissecans the femoral head. Clinical Orthopaedics and Related Research. 2010;**468**(2):613-618

[126] Aubin PP, et al. Long-term followup of fresh femoral osteochondral allografts for posttraumatic knee defects. Clinical Orthopaedics and Related Research. 2001;**391**(Suppl):S318-S327

[127] Kim CW, et al. Treatment of post-traumatic ankle arthrosis with bipolar tibiotalar osteochondral shell allografts. Foot & Ankle International. 2002;**23**(12):1091-1102

[128] Bugbee WD. Fresh osteochondral allografts. Journal of Knee Surgery. 2002;**15**(3):191-195

[129] Chu CR, et al. Articular cartilage transplantation. Clinical results in the knee. Clinical Orthopaedics and Related Research. 1999;**360**:159-168

[130] Ghazavi MT, et al. Fresh osteochondral allografts for post-traumatic osteochondral defects of the knee. Journal of Bone & Joint Surgery – British Volume. 1997;**79**(6): 1008-1013

[131] Bugbee WD, Convery FR. Osteochondral allograft transplantation. Clinics in Sports Medicine. 1999;**18**(1):67-75

[132] Emmerson BC, et al. Fresh osteochondral allografting in the treatment of osteochondritis dissecans of the femoral condyle. The American Journal of Sports Medicine. 2007;**35**(6):907-914

[133] Gross AE, et al. Fresh osteochondral allografts for posttraumatic knee defects: Long-term followup. Clinical Orthopaedics and Related Research. 2008;**466**(8):1863-1870

[134] Brittberg M, et al. Treatment of deep cartilage defects in the knee with autologous chondrocyte transplantation. New England Journal of Medicine. 1994;**331**(14):889-895

[135] Wood JJ, et al. Autologous cultured chondrocytes: Adverse events reported to the United States Food and Drug Administration. Journal of Bone & Joint Surgery – American Volume. 2006;**88**(3):503-507

[136] Peterson L, et al. Two- to 9-year outcome after autologous chondrocyte transplantation of the knee. Clinical Orthopaedics and Related Research, 2000;**374**:212-234

[137] Awad HA, et al. Chondrogenic differentiation of adipose-derived adult stem cells in agarose, alginate, and gelatin scaffolds. Biomaterials. 2004;**25**(16):3211-3222

[138] Guo JF, Jourdian GW, MacCallum DK. Culture and growth characteristics of chondrocytes encapsulated in alginate beads. Connective Tissue Research. 1989;**19**(2-4):277-297

[139] Pettersson S, et al. Cell expansion of human articular chondrocytes on macroporous gelatine scaffolds-impact of microcarrier selection on cell proliferation. Biomedical Materials. 2011;**6**(6):065001

[140] Marmotti A, et al. One-step osteochondral repair with cartilage fragments in a composite scaffold. Knee Surgery, Sports Traumatology, Arthroscopy. 2012;**20**(12):2590-2601

[141] Behrens P, et al. Matrix-associated autologous chondrocyte transplantation/implantation (MACT/MACI) – 5-year follow-up. Knee. 2006;**13**(3):194-202

[142] Marcacci M, et al. In: Williams RJ, editor, Cell-Based Cartilage Repair Using the Hyalograft Transplant. Cartilage Repair Strategies Totowa, NJ: Humana Press; 2007. pp. 207-218

[143] Benthien JP, Behrens P. Autologous Matrix-Induced Chondrogenesis (AMIC): Combining microfracturing and a collagen I/III matrix for articular cartilage resurfacing. Cartilage. 2010;**1**(1):65-68

[144] Gille J, et al. Outcome of Autologous Matrix-Induced Chondrogenesis (AMIC) in cartilage knee surgery: Data of the AMIC registry. Archives of Orthopaedic and Trauma Surgery. 2013;**133**(1):87-93

Chapter 7

Osteochondritis Dissecans of the Knee

Anthony C. Egger and Paul Saluan

Additional information is available at the end of the chapter

http://dx.doi.org/10.5772/intechopen.70275

Abstract

Osteochondritis dissecans (OCD) is a common but poorly understood source of knee pain and dysfunction. It is a condition primarily affecting the subchondral bone, with secondary effects on the articular cartilage surface. A large amount of research over the past two decades has produced many valuable insights into the condition, but further study and elucidation are still needed. The goal of this chapter will be to serve as a general overview of osteochondritis dissecans as it is understood today, including the etiology, clinical presentation, diagnosis, treatment options, outcomes, and future research aims.

Keywords: osteochondritis dissecans, knee, cartilage injury, OCD treatment, OCD outcomes

1. Introduction

Osteochondritis dissecans (OCD) has become a well-recognized, but still poorly understood source of knee pain and dysfunction. It is a condition primarily affecting the subchondral bone, with secondary effects on the articular cartilage surface. A large amount of research over the last two decades has produced many valuable insights into the condition, but further study and elucidation are still needed. The goal of this chapter will be to serve as a general overview of osteochondritis dissecans as it is understood today, including the etiology, clinical presentation, diagnosis, treatment options, outcomes, and future research aims.

The term osteochondritis dissecans was first documented in the literature in 1887 by Franz Konig, who described a presumed inflammatory process leading to loose bodies in the elbow and knee joints in young, atraumatic patients [1]. This theory was ultimately disproved as histological studies began to support findings of necrosis rather than inflammation in OCD lesions [2–6]. Many other theories and descriptions of osteochondritis dissecans have subsequently been proposed, but a definitive understanding remains elusive. The current working definition of OCD as developed by the leading collaborative research group on the topic is as follows: a focal, idiopathic alteration of subchondral bone with risk for instability and disruption of adjacent articular cartilage that result in premature osteoarthritis [7].

Traditionally, OCD had been subclassified into two groups based on the status of the distal femoral physis. Juvenile OCD occurs in those with an open distal femoral physis, whereas adult OCD is found in skeletally mature patients [8]. Previously, the etiology of OCD in skeletally immature individuals was thought to be from a fundamental disturbance in epiphyseal development. The adult form, on the other hand, was believed to be associated with more direct traumatic causation [9]. However, many experts now currently feel adult OCD is in the majority of cases not a distinct entity, but instead the natural progression of juvenile OCD missed in adolescence [10–12]. While the nomenclature is no longer as critical, the distinction between "juvenile" and "adult" OCD as based on presentation and timing of diagnosis is still important in regards to prognosis. Multiple studies have shown that juvenile OCD lesions to be more stable in appearance and to have a better prognosis than those diagnosed in adulthood [8, 10, 13–15].

2. Epidemiology

The presence of articular cartilage pathology is found in greater than 60% of patients undergoing knee arthroscopies, with focal chondral defects of all varieties found in 20% of these patients [16–18]. As a subset of these lesions, osteochondritis dissecans of the knee remains a relatively uncommon condition. In the pediatric population aged 6–19 years, the incidence of OCD lesions of the knee was found to be 9.5 per 100,000. There is a strong predilection for males versus females with an incidence of 15.4 and 3.3 per 100,000, respectively. Patients aged 12–19 years have an over three times risk of OCD than those aged 6–11 years. In terms of race and ethnicity, African Americans have double the risk of OCD of the knee compared to non-Hispanic whites, and at least 4 times the risk of disease as all other races and ethnicities [19].

The most common location for OCD lesions to occur is in the medial femoral condyle, which accounts for 70–85% of all lesions. The majority of these lesions occur in the posterolateral aspect of the medial femoral condyle [19]. The next most frequent location is the lateral femoral condyle, and the lesions in this location are often found to be larger and more advanced. OCD lesions are also rarely found on the patella, trochlea, and tibial plateau [11].

3. Etiology

Starting with Konig's inflammatory theory, numerous hypotheses regarding the true pathophysiology behind the formation and progression of OCD lesions of the knee have been proposed, but no one theory has gained uniform consensus. In histologic review of OCD lesions, necrosis of the subchondral bone is often identified but it remains unclear if the presence of the necrosis is primary or secondary to the pathogenesis of OCD [3–6, 20]. The vascularity of the subchondral bone has been described as an end arterial arcade with poor anastomoses.

Histology of necrotic bone has been shown to be consistent with vascular occlusions, and it has been proposed that insufficient arterial branching could lead to subchondral bone infarction and subsequent OCD [21, 22]. The presence of an ischemic zone in the lateral aspect of the medial femoral condyle has been questioned, although particularly in young patients who have good distal femoral blood supply [23–26].

Repetitive microtrauma has become the most accepted cause of OCD, mainly due to the rising incidence of the disorder among athletes [27]. The theory states that an initial stress reaction occurs in the subchondral bone of the knee and with further loading a true stress fracture is generated. Repetitive, progressive loading prevents the stress fracture from healing and eventually the subchondral bone becomes necrotic [2]. The fragment begins to dissect and ultimately separate from the fracture bed leading to an unstable OCD lesion. In this theory, bone necrosis is seen as secondary to trauma rather than to a primary lack of vascularity. Mechanical axis alignment has also been associated with OCD, with aberrant mechanical pressures on the condyles potentially leading to the formation of an OCD [28]. The true etiology of OCD is most likely multi-factorial and a combination of the currently proposed theories.

4. Clinical presentation and physical examination

The clinical presentation of OCD lesions can be quite variable and often differs depending on the stability and severity of the lesion. Stable lesions, as are frequently seen in juvenile OCD, often present with complaints of nonspecific and poorly localized knee pain which is exacerbated by exercise, particularly when climbing stairs or hills [10]. Unstable lesions are commonly seen in adult OCD and present with more mechanical symptoms like swelling, stiffness, locking, and catching.

On physical examination, both stable and unstable lesions may present with an antalgic gait. An external rotation of the tibia during gait can be seen as compensation for impingement of the tibial eminence on an OCD lesion of the medial femoral condyle [29]. This can be tested clinically with the Wilson test, which elicits pain when the tibia is internally rotated during extension of the knee between 90 and 30°. Pain is relieved with tibial external rotation as it moves the eminence away from the lesion. Ligamentous stability and overall alignment must also be assessed to allow for concomitant pathology to be appropriately addressed. Muscle strength testing is also important as significant dynamic strength deficits of the quadriceps and core may warrant rehabilitation attempts prior to surgery [30].

5. Imaging

Plain radiographs and magnetic resonance imaging (MRI) are the two most commonly used imaging modalities in evaluating knee OCD. Radiographs are commonly used for the initial diagnosis and assessment of skeletal maturity, whereas MRI highlights changes in the articular cartilage and subchondral bone.

Radiographs are relatively inexpensive and easy to obtain, making it the initial imaging choice for evaluation of suspected OCD. Radiographs evaluating for OCD lesions often include anteroposterior, lateral, sunrise, and tunnel/notch views. The characteristic appearance of an OCD lesion of the knee consists of a well-circumscribed lucent defect in the subchondral bone [31]. The notch view, which is obtained with a posterior to anterior beam at approximately 30° of flexion, is particularly helpful for evaluating the posterior aspects of the femoral condyles [8]. Evaluating for potential lesions in boys younger than 13 and girls younger than 11 requires caution as they may develop secondary ossifications that can resemble OCD lesions and MRI is often needed for clarification [10]. Given the limitations of radiographs in assessing an OCD lesion, MRI is often used to evaluate the true size and stability in order to determine an appropriate surgical plan.

6. Classification systems

The most commonly used classification system for OCD lesions is based on MRI findings. The Hefti system divides lesions into five different stages and differentiates between stable (stages 1 and 2) and unstable (stages 3, 4, and 5) lesions with progressive pathology noted [13].

The MRI classification has been shown to be accurate to divide lesions into stable and unstable categories, but ultimately arthroscopic evaluation provides the best assessment of the OCD lesion [32]. Multiple arthroscopic systems have been proposed to classify lesions during surgery, but no comprehensive system to describe the full complement of OCD lesions has been accepted [32–37]. The Research in OsteoChondritis of the Knee (ROCK) group developed a novel classification system to provide a common language in describing these lesions [38]. To optimize comprehensibility and applicability, each type was described with a memorable name. The classification divides lesions into immobile and mobile lesions. The "cue ball" (no detectable abnormality), "shadow" (cartilage intact but subtly demarcated), and "wrinkle in the rug" (cartilage is demarcated with a fissure or wrinkle) are in the immobile category. The mobile lesions consist of the "locked door" (cartilage fissuring at periphery but unable to hinge open), "trap door" (able to hinge open the fissure), and "crater" (exposed subchondral bone defect). This classification system has been shown to have very good inter-observer reliability and should be used to facilitate a common language which is crucial for future collaborative research.

7. Nonoperative treatment

Nonoperative management is the appropriate first line of treatment for stable juvenile OCD lesions. Juvenile OCD lesions have a higher healing potential than adult lesions, and an open distal femoral physis has been shown as one of the best predictors for successful nonoperative management [39]. Conservative management is usually attempted for a minimum of 3 months to allow for potential healing. Most current nonoperative treatment plans focus on activity modification with cessation of impact activities and protected weight bearing with crutches or an offloader brace [40]. The goal of conservative management is to eliminate pain and repetitive loading to help promote healing of OCD lesions.

Overall, successful healing rates >50% have been shown for stable juvenile OCD lesions treated nonoperatively. However, this has not been replicated in the adult population, with poor results seen without surgical intervention for adult OCD lesions [4, 8, 41]. Adult OCD lesions have little capacity for healing with nonoperative means, but an unloader brace is a potential temporary option to allow an athlete to finish their season prior to operative intervention [40]. Complete resolution of symptoms takes time, patience, and compliance, which is important to stress to patients early in the process.

8. Operative treatment

In those patients who have failed nonoperative treatment or have large, unstable, or unsalvageable lesions, surgical intervention is often required. Cartilage treatment strategies can be characterized as palliation (debridement), repair [drilling and microfracture (MF)], or restoration [osteochondral autograft transfer (OAT), osteochondral allograft (OCA), and autologous chondrocyte implantation (ACI)] [42]. One of the most important determinations to be made prior to surgical intervention is the stability of the OCD lesion. The stability relates to the mechanical integrity of the subchondral lesion [43]. A lesion which is immobile and resting in situ is considered to be stable, whereas a lesion which is mobile, fragmented, or ex situ is considered unstable. The distinction is important for determining the appropriate surgical plan.

9. Subchondral drilling

Subchondral drilling is the initial standard of care operative procedure for stable OCD lesions. There are two main types of drilling, transarticular and retroarticular, but the principle behind each technique is the same. The goal of subchondral drilling is to use a Kirschner wire to disrupt the sclerotic margin of the lesion to establish channels between the necrotic subchondral bone and the healthy cancellous bone in order to promote revascularization, osseous bridging, and healing [11]. The average time to healing is around 4–6 months after surgery.

Transarticular drilling is done from inside the joint and penetrates the articular cartilage through at least one site to create subchondral penetrations. The main concern with this technique is related to the uncertain long-term implications of disrupting the articular cartilage with the drill sites. Retroarticular drilling avoids this concern by sparing the articular surface and physes with drilling through the affected femoral condyle into the lesion under fluoroscopy. Aside from the added radiation risk, this technique is also more technically demanding and risks incomplete lesion drilling, lesion displacement, or inadvertent soft tissue injury [10].

Neither technique has clearly demonstrated superior patient-orientated outcomes or radiographic healing. Transarticular drilling demonstrated an average healing rate of 91% with a mean healing time of 4.5 months with retroarticular just behind at 86% at 5.6 months [44].

No complications were noted throughout a review of all studies on drilling, making the technique not only effective but also safe option. Poorer results have been noted in older patients with closed physes, fissures of the articular cartilage, and lesions located outside the traditional posterolateral medial condyle [45–47].

10. Debridement

The simplest solution to the management of an unstable OCD lesion is excision of the fragment with debridement of the remaining chondral defect. As the painful and limiting mechanical symptoms of an unstable OCD are due to these loose fragments, excision has been correlated with good short-term clinical results [48, 49]. However, as excision and debridement alone leads to a loss of articular cartilage with subsequent degenerative changes, the longer term imaging and knee function scores deteriorate [50, 51]. Even while patients maintain good clinical knee scores, evidence of early degenerative changes can be seen on radiographs at midterm follow up after excision and debridement [52, 53]. The results of these studies further reinforce that every attempt should be made to preserve, repair, or replace the native bone and cartilage that is damaged in an OCD lesion.

11. Lesion fixation

For unstable lesions or stable lesions that have failed a drilling procedure, the next surgical option is often fixation of the osteochondral lesion. The general principles of lesion fixation are to attempt to restore the articular surface, enhance the blood supply of the osseous interface, and initiate early range of motion postoperatively [8].

Historically, after lesions were debrided and bone grafted, they were pinned in place with Kirschner wires; after, the lesion had been debrided and bone grafting had been applied [54]. However, this technique has largely been abandoned due to K-wire bending and inability to hold and provide an adequate compressive force to the lesion. K-wires were replaced by the use of rigid metal screw fixation, either with variable pitch or cannulated partially threaded screws. Most recently, bioabsorbable implants designed as screws or pins have become popularized for fixation. Fixation is again particularly important given the poor results seen with detached fragment removal, especially in weight-bearing areas of the femoral condyles [49, 51, 55].

Variable pitch headless screws were initially described for use in scaphoid fixation, but indications spread to include fixation of OCD lesions [56–58]. The goal of fixation with these screws is to achieve compression encouraging bony union of the subchondral fractures. The main advantage of variable pitch headless screws (Herbert screws) lies in their ability to provide strong compression and be sunk completely under the articular surface to prevent protrusion. The rigid fixation also allows early joint motion due to anatomic restoration of the joint surface [59]. The majority of patients undergoing this technique report good to excellent results without major complications [60–62]. The use of cannulated screws has also been described with successful results; however, the major drawback is the concern for increased articular cartilage morbidity due to screw prominence on the articular surface [63, 64]. Cannulated headless compression screws have now been developed as an alternative, which theoretically combines the advantages of both techniques [65].

Bioabsorbable screws, pins, tacks, and darts have been designed and utilized with overall good results [66–69]. The main advantages of bioabsorbable fixation are the lack of metal artifact on postoperative MRI as well as theoretically no subsequent surgery needed for implant removal [66]. Bioabsorbable implants can fail though due to screw breakage, screw back out, reactive synovitis, and loss of compressive force over time [70–73]. These implant failures often lead to refractory mechanical symptoms and need for revision surgery. Despite these potential risks, unstable lesions should still be fixed instead of excised when technically feasible. As there has been no significant difference noted in comparison of bioabsorbable pins and tacks, variable pitch screws, and partially threaded screws with regard to clinical and radiographic healing, the choice of fixation is surgeon dependent [69]. The most frequently used techniques among surgeons are bioabsorbable screws and metal headless variable pitch screws [74].

12. Microfracture

Microfracture is a marrow stimulation technique that was developed and implemented in the early 1980s to allow for cartilage repair. The goal of the procedure is to create microfractures in the subchondral bone perpendicular to the surface to create a surface rough enough to hold the generated marrow clot. The pluripotent cells of the clot proliferate and differentiate into cells with morphological features similar to chondrocytes. These cells then produce a cartilaginous repair tissue to fill the chondral defect [75]. The fibrocartilage which matures though is often predominately type 1 collagen, a structurally different entity from hyaline cartilage [76].

Indications for microfracture include smaller partial and full-thickness cartilage defects in patients with acceptable knee alignment. The greatest improvement occurs with the treatment of acute lesions less than 4 cm in size in patients under 35 years old [77]. Younger patients have better results with microfracture as it is crucial to have adequate height of cartilage on the lesion rim to hold the clot in place, which is difficult in degenerative lesions where the cartilage is thinner [75].

Early results of microfracture are positive with clear improvement in knee function noted throughout the literature at 2 years, particularly in smaller lesions. Despite good midterm results published by the developer of the technique, the longevity and durability of microfracture have been questioned [78–80]. When compared to other cartilage procedures like OAT and ACI, the results are mixed, although no study showed superior results for microfracture [81–83]. Microfracture has been found to have a significantly higher failure rate and need for reoperation than OAT or ACI with larger lesions (>4.5 cm^2) and at greater than 5 years postoperatively [84]. An even smaller size threshold (<2 cm^2) has been shown for microfracture to be successful in the demanding athletic population [81, 85].

13. Autograft transplantation

In the cases of failed fixation, lesion fragmentation, or chronically detached lesions, more advanced chondral procedures, like osteochondral autograft transplantation (OAT), are required. OAT was developed and then popularized in the 1990s [86, 87]. The procedure entails the harvesting of a cylindrical graft of healthy cartilage and subchondral bone from a less stressed area of the distal femur and implementing into an area of chondral defect. The graft is matched to the surface area of the defect and seated to restore a smooth cartilage surface in the joint [88]. A single plug of cartilage may be transferred or an alternative procedure termed mosaicplasty can be performed where multiple smaller plugs are implemented.

Osteochondral autograft transplantation is currently recommended as a viable option for osteochondral lesions measuring 1–4 cm^2 in a load-bearing area [89]. OAT offers the opportunity to repair cartilaginous defects by restoring hyaline cartilage anatomy [90]. Graft plugs should be taken from nonweight-bearing areas to avoid being arthrogenic [91]. OAT provides an immediate functional surface that allows a relatively quick rehabilitation and return to play, but a mismatch of cartilage thickness between the two sites can lead to abnormal stresses and poor function [92, 93].

Mosaicplasty has been shown to give reliably good short-term results [94–97]. In longer term studies evaluating patients who underwent mosaicplasty, there is a significant decrease in level of physical activity noted, particularly in patients whose activity level prior to surgery was high. This reduction in activity level is often due to apprehension and a desire to preserve the joint [91]. Older age, female sex, and more extensive initial lesions have been shown to be factors leading to poor prognoses after mosaicplasty [98]. Limb malalignment has also been shown to affect outcomes if not corrected, and thus concomitant osteotomy is recommended in these cases [99].

The primary concern with autografting comes from possible donor site morbidity. Cadaveric studies have shown load across donor sites during range of motion, but multiple studies have shown minimal to no complications associated with donor sites at midterm follow-up [81, 95, 97, 100]. Athletes report nearly double the rate of donor site pain compared to less active patients, indicating that vigorous exercise potentially increases donor site pain [99].

14. Allograft transplantation

Osteochondral allograft transplantation (OCA) involves the transfer of size-matched allograft cartilage and subchondral bone into large osteochondral defects of the knee [30]. OCA is primarily used in the management of large osteochondral defects and as a salvage option for those who have previously failed other cartilage repair techniques. Fresh osteochondral allograft transplantation is theoretically an attractive option because it can restore both the osseous and the chondral components caused by the OCD lesion [101].

Allograft tissue is harvested within 24 hours of donor death, ideally from a donor aged 15–40 years with grossly healthy articular cartilage [102]. Allografts are often matched by tissue banks based on size, which is usually measured off an AP radiograph of the knee. The affected condyle is used for sizing and a match is sought based on the overall condyle size, with an acceptable match noted to be within ±2 mm. While it is preferred to have patient size, side, and condyle-specific matching, depending on the location of the lesion, it has been shown that plugs may be successfully transplanted to the other compartment (medial to lateral) or even to the other side (left vs. right). Once harvested, OCAs should be properly stored and implanted within 28 days for maintained chondrocyte viability and subsequent clinical benefit [103–108].

OCA is effective as a majority of patients are satisfied with their treatment and are able to return to sport or recreational activity [109]. The success of OCA is highest when a single articular surface is replaced, the surrounding ligaments and menisci are intact, and the alignment is normal [110]. Osteoarthritis or the presence of disease on both articular surfaces is a contraindication to OCA [111]. The number of previous ipsilateral knee surgical procedures, elevated BMI, age >30 years old, and medial femoral graft location have been found to be independent factors predictive of reoperation and failure after allograft transplantation [101, 112].

Overall, there is a 1 in 3 chance of undergoing an additional operation, with vast majority being arthroscopic debridement, within the first 5 years following OCA. Despite this high rate of requiring a second surgery, OCA remains an attractive option due to allograft having the ability to treat larger defects, the lack of donor site morbidity, reduced surgical time, and the ability to customize the graft to the recipient's defect site.

15. Autologous chondrocyte implantation

Autologous chondrocyte implantation is a two-stage procedure indicated for full thickness cartilage or OCD lesions of the knee. The initial procedure involves arthroscopic evaluation and cartilage harvesting. After 2 weeks of culturing, the harvested chondrocytes are then implanted and sealed into the cartilage defect in an attempt to recreate a hyaline cartilage interface. ACI is indicated for full thickness cartilage or osteochondral lesions of the knee ranging from 2 to 16 cm^2 with minimal cartilage damage on the opposing articular surface [113].

The treatment of OCD lesions with ACI has been associated with clinical improvements, including reduced pain and improved function, in both adolescents and adults at midterm follow-up [114–117]. As with other cartilage repair techniques, younger patients with more localized lesions tend to do better [118–120].

A drawback to ACI is the requirement of two separate procedures. However, most patients undergoing ACI have already failed numerous other options and are willing to undergo the extra surgery for a chance at salvage. Most complications of ACI seem to be related to the periosteal flap, including overgrowth, delamination, and arthrofibrosis. Majority of failures occur with the first 2 years after surgery [121]. Despite these limitations, ACI remains a cartilage salvage option, particularly in those who have failed other surgical modalities.

16. Future research

Despite over 100 years of research, there is still much to be learned regarding osteochondritis dissecans. In 2011, the American Academy of Orthopedic Surgeons released Clinical Practice guidelines regarding OCD of the knee [122]. These guidelines found limited evidence for all aspects of the treatment of knee OCD. To provide better insight and advance the understanding of this condition, multicenter study research groups have been formed. These groups are undertaking clinical trials attempting to answer many of the unsolved issues relating to knee OCD [123].

17. Conclusion

Osteochondritis dissecans of the knee remains a poorly understood and difficult problem-facing patients and orthopedic surgeons today. Affecting both articular cartilage and subchondral bone, OCD is a progressive condition leading to knee pain, mechanical symptoms, and ultimately osteoarthritis if left untreated. OCD recognized in patients with open distal femoral physes is termed juvenile OCD and has a better prognosis, particularly with nonoperative management. Adult OCD is found in patients after skeletal maturity and almost always requires surgical intervention. The stability and size of the lesion is critical in determining the appropriate surgical modality. Reparative procedures such as drilling, microfracture, and lesion stabilization have shown good early results for smaller lesions, but larger and more chronic lesions often require regenerative chondral techniques like osteochondral autograft, allograft, or acellular chondrocyte implantation. Further research is underway comparing the different techniques to determine the gold standard for each size and type of lesion. The interest and understanding of knee OCD has progressed considerably in the past 20 years, but still more prospective research studies are needed to improve the assessment and treatment of this complex condition.

Author details

Anthony C. Egger and Paul Saluan*

*Address all correspondence to: saluanp@ccf.org

Department of Orthopaedics, The Cleveland Clinic Foundation, Cleveland, OH, United States

References

[1] König F. The classic: On loose bodies in the joint. Clinical Orthopaedics and Related Research®. 2013;**471**(4):1107-1115

[2] Shea KG, Jacobs JC, Carey JL, Anderson AF, Oxford JT. Osteochondritis dissecans knee histology studies have variable findings and theories of etiology. Clinical Orthopaedics and Related Research®. 2013;**471**(4):1127-1136

[3] Campbell CJ, Ranawat CS. Osteochondritis dissecans: The question of etiology. Journal of Trauma and Acute Care Surgery. 1966;**6**(2):201-221

[4] Linden B, Telhag H. Osteochondritis dissecans: A histologic and autoradiographic study in man. Acta Orthopaedica Scandinavica. 1977;**48**(6):682-686

[5] Portigliatti Barbos M, Brach del Prever E, Borroni L, Salvadori L, Battiston B. Osteochondritis dissecans of the femoral condyles. A histological study with pre-operative fluorescent bone labelling and microradiography. Italian Journal of Orthopaedics and Traumatology. 1985;**11**(2):207-213

[6] Uozumi H, Sugita T, Aizawa T, Takahashi A, Ohnuma M, Itoi E. Histologic findings and possible causes of osteochondritis dissecans of the knee. American Journal of Sports Medicine. 2009;**37**(10):2003-2008

[7] Edmonds EW, Shea KG. Osteochondritis dissecans: Editorial comment. Clinical Orthopaedics and Related Research. 2013;**471**(4):1105

[8] Cahill BR. Osteochondritis dissecans of the knee: Treatment of juvenile and adult forms. Journal of the American Academy of Orthopaedic Surgeons. 1995;**3**(4):237-247

[9] Smillie IS. Osteochondritis Dissecans Loose Bodies in Joints. London: E. & S. Livingstone Ltd.; 1960

[10] Kocher MS, Tucker R, Ganley TJ, Flynn JM. Management of osteochondritis dissecans of the knee: Current concepts review. American Journal of Sports Medicine. 2006;**34**(7):1181-1191

[11] Heyworth BE, Kocher MS. Osteochondritis dissecans of the knee. JBJS Reviews. 2015;**3**(7):e1

[12] Edmonds EW, Polousky J. A review of knowledge in osteochondritis dissecans: 123 years of minimal evolution from König to the ROCK study group. Clinical Orthopaedics and Related Research®. 2013;**471**(4):1118-1126

[13] Hefti F, Beguiristain J, Krauspe R, et al. Osteochondritis dissecans: A multicenter study of the European pediatric orthopedic society. Journal of Pediatric Orthopaedics B. 1999;**8**(4):231-245

[14] Cepero S, Ullot R, Sastre S. Osteochondritis of the femoral condyles in children and adolescents: Our experience over the last 28 years. Journal of Pediatric Orthopaedics B. 2005;**14**(1):24-29

[15] Bradley J, Dandy DJ. Osteochondritis dissecans and other lesions of the femoral condyles. Journal of Bone and Joint Surgery. British. 1989;**71**(3):518-522

[16] Aroen A, Loken S, Heir S, et al. Articular cartilage lesions in 993 consecutive knee arthroscopies. American Journal of Sports Medicine. 2004;**32**(1):211-215

[17] Widuchowski W, Widuchowski J, Trzaska T. Articular cartilage defects: Study of 25,124 knee arthroscopies. The Knee. 2007;**14**(3):177-182

[18] Hjelle K, Solheim E, Strand T, Muri R, Brittberg M. Articular cartilage defects in 1,000 knee arthroscopies. Arthroscopy: The Journal of Arthroscopic & Related Surgery. 2002;**18**(7):730-734

[19] Kessler JI, Nikizad H, Shea KG, Jacobs Jr JC, Bebchuk JD, Weiss JM. The demographics and epidemiology of osteochondritis dissecans of the knee in children and adolescents. American Journal of Sports Medicine. 2014;**42**(2):320-326

[20] Yonetani Y, Matsuo T, Nakamura N, et al. Fixation of detached osteochondritis dissecans lesions with bioabsorbable pins: Clinical and histologic evaluation. Arthroscopy: The Journal of Arthroscopic & Related Surgery. 2010;**26**(6):782-789

[21] Schenck RC Jr., Goodnight JM. Osteochondritis dissecans. Journal of Bone and Joint Surgery. American. 1996;**78**(3):439-456

[22] Enneking WF. Clinical Musculoskeletal Pathology. University of Florida Press/J. Hillis Miller Health Science Center; Gainseville, Florida. 1990

[23] Koch S, Kampen W, Laprell H. Cartilage and bone morphology in osteochondritis dissecans. Knee Surgery, Sports Traumatology, Arthroscopy. 1997;**5**(1):42-45

[24] Reddy AS, Frederick RW. Evaluation of the intraosseous and extraosseous blood supply to the distal femoral condyles. American Journal of Sports Medicine. 1998;**26**(3):415-419

[25] Rogers WM, Gladstone H. Vascular foramina and arterial supply of the distal end of the femur. Journal of Bone and Joint Surgery. America. 1950;**32 A**(4):867-874

[26] Wall E, Von Stein D. Juvenile osteochondritis dissecans. Orthopedic Clinics of North America. 2003;**34**(3):341-353

[27] Gornitzky AL, Mistovich RJ, Atuahuene B, Storey EP, Ganley TJ. Osteochondritis dissecans lesions in family members: Does a positive family history impact phenotypic potency? Clinical Orthopaedics and Related Research®. June 2017:**475**(6);1573-1580

[28] Hughston JC, Hergenroeder PT, Courtenay BG. Osteochondritis dissecans of the femoral condyles. Journal of Bone and Joint Surgery. America. 1984;**66**(9):1340-1348

[29] Wilson J. A diagnostic sign in osteochondritis dissecans of the knee. JBJS. 1967;**49**(3):477-480

[30] Sherman SL, Garrity J, Bauer K, Cook J, Stannard J, Bugbee W. Fresh osteochondral allograft transplantation for the knee: Current concepts. Journal of the American Academy of Orthopaedic Surgeons. 2014;**22**(2):121-133

[31] Zbojniewicz AM, Laor T. Imaging of osteochondritis dissecans. Clinics in Sports Medicine. 2014;**33**(2):221-250

[32] O'Connor MA, Palaniappan M, Khan N, Bruce CE. Osteochondritis dissecans of the knee in children. A comparison of MRI and arthroscopic findings. Journal of Bone and Joint Surgery. British. 2002;**84**(2):258-262

[33] Chen C, Liu Y, Chou P, Hsieh C, Wang C. MR grading system of osteochondritis dissecans lesions: Comparison with arthroscopy. European Journal of Radiology. 2013;**82**(3):518-525

[34] Nelson DW, DiPaola J, Colville M, Schmidgall J. Osteochondritis dissecans of the talus and knee: Prospective comparison of MR and arthroscopic classifications. Journal of Computer Assisted Tomography. 1990;**14**(5):804-808

[35] Jacobs JC, Archibald-Seiffer N, Grimm NL, Carey JL, Shea KG. A review of arthroscopic classification systems for osteochondritis dissecans of the knee. Orthopedic Clinics of North America. 2015;**46**(1):133-139

[36] Brittberg M, Winalski CS. Evaluation of cartilage injuries and repair. Journal of Bone and Joint Surgery. America. 2003;**85-A**(2):58-69

[37] Dipaola JD, Nelson DW, Colville MR. Characterizing osteochondral lesions by magnetic resonance imaging. Arthroscopy: The Journal of Arthroscopic & Related Surgery. 1991;**7**(1):101-104

[38] Carey JL, Wall EJ, Grimm NL, et al. Novel arthroscopic classification of osteochondritis dissecans of the knee: A multicenter reliability study. American Journal of Sports Medicine. 2016;**44**(7):1694-1698

[39] Paletta GA Jr., Bednarz PA, Stanitski CL, Sandman GA, Stanitski DF, Kottamasu S. The prognostic value of quantitative bone scan in knee osteochondritis dissecans. A preliminary experience. American Journal of Sports Medicine. 1998;**26**(1):7-14

[40] Carey JL, Grimm NL. Treatment algorithm for osteochondritis dissecans of the knee. Clinics in Sports Medicine. 2014;**33**(2):375-382

[41] DellaMaggiora R, Vaishnav S, Vangsness CT. Osteochondritis dissecans of the adult knee. Operative Techniques in Sports Medicine. 2008;**16**(2):65-69

[42] McNickle AG, Provencher MT, Cole BJ. Overview of existing cartilage repair technology. Sports Medicine and Arthroscopy Review. 2008;**16**(4):196-201

[43] Mesgarzadeh M, Sapega AA, Bonakdarpour A, et al. Osteochondritis dissecans: Analysis of mechanical stability with radiography, scintigraphy, and MR imaging. Radiology. 1987;**165**(3):775-780

[44] Gunton MJ, Carey JL, Shaw CR, Murnaghan ML. Drilling juvenile osteochondritis dissecans: Retro-or transarticular? Clinical Orthopaedics and Related Research®. 2013;**471**(4):1144-1151

[45] Boughanem J, Riaz R, Patel RM, Sarwark JF. Functional and radiographic outcomes of juvenile osteochondritis dissecans of the knee treated with extra-articular retrograde drilling. American Journal of Sports Medicine. 2011;**39**(10):2212-2217

[46] Kocher MS, Micheli LJ, Yaniv M, Zurakowski D, Ames A, Adrignolo AA. Functional and radiographic outcome of juvenile osteochondritis dissecans of the knee treated with transarticular arthroscopic drilling. American Journal of Sports Medicine. 2001;**29**(5):562-566

[47] Louisia S, Beaufils P, Katabi M, Robert H. Transchondral drilling for osteochondritis dissecans of the medial condyle of the knee. Knee Surgery, Sports Traumatology, Arthroscopy. 2003;**11**(1):33-39

[48] Denoncourt PM, Patel D, Dimakopoulos P. Arthroscopy update #1. treatment of osteochondrosis dissecans of the knee by arthroscopic curettage, follow-up study. Orthopedic Reviews. 1986;**15**(10):652-657

[49] Aglietti P, Ciardullo A, Giron F, Ponteggia F. Results of arthroscopic excision of the fragment in the treatment of osteochondritis dissecans of the knee. Arthroscopy: The Journal of Arthroscopic & Related Surgery. 2001;**17**(7):741-746

[50] Michael J, Wurth A, Eysel P, König D. Long-term results after operative treatment of osteochondritis dissecans of the knee joint—30 year results. International Orthopaedics. 2008;**32**(2):217-221

[51] Anderson AF, Pagnani MJ. Osteochondritis dissecans of the femoral condyles. Long-term results of excision of the fragment. American Journal of Sports Medicine. 1997;**25**(6):830-834

[52] Wright RW, McLean M, Matava MJ, Shively RA. Osteochondritis dissecans of the knee: Long-term results of excision of the fragment. Clinical Orthopaedics and Related Research®. 2004;**424**:239-243

[53] Murray J, Chitnavis J, Dixon P, et al. Osteochondritis dissecans of the knee; long-term clinical outcome following arthroscopic debridement. The Knee. 2007;**14**(2):94-98

[54] Anderson AF, Lipscomb AB, Coulam C. Antegrade curettement, bone grafting and pinning of osteochondritis dissecans in the skeletally mature knee. American Journal of Sports Medicine. 1990;**18**(3):254-261

[55] Sanders TL, Pareek A, Obey MR, et al. High rate of osteoarthritis after osteochondritis dissecans fragment excision compared with surgical restoration at a mean 16-year follow-up. American Journal of Sports Medicine. 2017;**45**(8):1799-1805

[56] Herbert TJ, Fisher WE. Management of the fractured scaphoid using a new bone screw. Journal of Bone and Joint Surgery. British. 1984;**66**(1):114-123

[57] Wombwell JH, Nunley JA. Compressive fixation of osteochondritis dissecans fragments with Herbert screws. Journal of Orthopaedic Trauma. 1987;**1**(1):74-77

[58] Thomson NL. Osteochondritis dissecans and osteochondral fragments managed by Herbert compression screw fixation. Clinical Orthopaedics and Related Research®. 1987;**224**:71-78

[59] Kouzelis A, Plessas S, Papadopoulos AX, Gliatis I, Lambiris E. Herbert screw fixation and reverse guided drillings, for treatment of types III and IV osteochondritis dissecans. Knee Surgery, Sports Traumatology, Arthroscopy. 2006;**14**(1):70-75

[60] Zuniga JR, Sagastibelza J, Blasco JL, Grande MM. Arthroscopic use of the Herbert screw in osteochondritis dissecans of the knee. Arthroscopy: The Journal of Arthroscopic & Related Surgery. 1993;**9**(6):668-670

[61] Mackie IG, Pemberton DJ, Maheson M. Arthroscopic use of the Herbert screw in osteochondritis dissecans. Journal of Bone and Joint Surgery. British. 1990;**72**(6):1076

[62] Makino A, Muscolo DL, Puigdevall M, Costa-Paz M, Ayerza M. Arthroscopic fixation of osteochondritis dissecans of the knee: Clinical, magnetic resonance imaging, and arthroscopic follow-up. American Journal of Sports Medicine. 2005;**33**(10):1499-1504

[63] Cugat R, Garcia M, Cusco X, et al. Osteochondritis dissecans: A historical review and its treatment with cannulated screws. Arthroscopy: The Journal of Arthroscopic & Related Surgery. 1993;**9**(6):675-684

[64] Johnson LL, Uitvlugt G, Austin MD, Detrisac DA, Johnson C. Osteochondritis dissecans of the knee: Arthroscopic compression screw fixation. Arthroscopy: The Journal of Arthroscopic & Related Surgery. 1990;**6**(3):179-189

[65] Barrett I, King AH, Riester S, et al. Internal fixation of unstable osteochondritis dissecans in the skeletally mature knee with metal screws. Cartilage. 2016;**7**(2):157-162

[66] Tabaddor RR, Banffy MB, Andersen JS, et al. Fixation of juvenile osteochondritis dissecans lesions of the knee using poly 96L/4D-lactide copolymer bioabsorbable implants. Journal of Pediatric Orthopaedics. 2010;**30**(1):14-20

[67] Camathias C, Gögüs U, Hirschmann MT, et al. Implant failure after biodegradable screw fixation in osteochondritis dissecans of the knee in skeletally immature patients. Arthroscopy: The Journal of Arthroscopic & Related Surgery. 2015;**31**(3):410-415

[68] Wouters DB, van Horn JR, Bos RR. The use of biodegradables in the treatment of osteochondritis dissecans of the knee: Fiction or future? Acta Orthopaedica Belgica. 2003;**69**(2):175-181

[69] Kocher MS, Czarnecki JJ, Andersen JS, Micheli LJ. Internal fixation of juvenile osteochondritis dissecans lesions of the knee. American Journal of Sports Medicine. 2007;**35**(5):712-718

[70] Friederichs MG, Greis PE, Burks RT. Pitfalls associated with fixation of osteochondritis dissecans fragments using bioabsorbable screws. Arthroscopy: The Journal of Arthroscopic & Related Surgery. 2001;**17**(5):542-545

[71] Scioscia TN, Giffin JR, Allen CR, Harner CD. Potential complication of bioabsorbable screw fixation for osteochondritis dissecans of the knee. Arthroscopy: The Journal of Arthroscopic & Related Surgery. 2001;**17**(2):1-5

[72] Fridén T, Rydholm U. Severe aseptic synovitis of the knee after biodegradable internal fixation: A case report. Acta Orthopaedica Scandinavica. 1992;**63**(1):94-97

[73] Barfod G, Svendsen RN. Synovitis of the knee after intraarticular fracture fixation with biofix®: Report of two cases. Acta Orthopaedica Scandinavica. 1992;**63**(6):680-681

[74] Yellin JL, Gans I, Carey JL, Shea KG, Ganley TJ. The surgical management of osteochondritis dissecans of the knee in the skeletally immature: A survey of the pediatric orthopaedic society of North America (POSNA) membership. Journal of Pediatric Orthopaedics. 2015

[75] Steadman JR, Rodkey WG, Rodrigo JJ. Microfracture: Surgical technique and rehabilitation to treat chondral defects. Clinical Orthopaedics and Related Research®. 2001;**391**:S362-S369

[76] Frisbie DD, Oxford JT, Southwood L, et al. Early events in cartilage repair after subchondral bone microfracture. Clinical Orthopaedics and Related Research®. 2003;**407**:215-227

[77] Steadman JR, Rodkey WG, Singleton SB, Briggs KK. Microfracture technique for full-thickness chondral defects: Technique and clinical results. Operative Techniques in Orthopaedics. 1997;**7**(4):300-304

[78] Steadman JR, Briggs KK, Rodrigo JJ, Kocher MS, Gill TJ, Rodkey WG. Outcomes of microfracture for traumatic chondral defects of the knee: Average 11-year follow-up. Arthroscopy: The Journal of Arthroscopic & Related Surgery. 2003;**19**(5):477-484

[79] Goyal D, Keyhani S, Lee EH, Hui JHP. Evidence-based status of microfracture technique: A systematic review of level I and II studies. Arthroscopy: The Journal of Arthroscopic & Related Surgery. 2013;**29**(9):1579-1588

[80] Mithoefer K, McAdams T, Williams RJ, Kreuz PC, Mandelbaum BR. Clinical efficacy of the microfracture technique for articular cartilage repair in the knee: An evidence-based systematic analysis. American Journal of Sports Medicine. 2009;**37**(10):2053-2063

[81] Gudas R, Kalesinskas RJ, Kimtys V, et al. A prospective randomized clinical study of mosaic osteochondral autologous transplantation versus microfracture for the treatment of osteochondral defects in the knee joint in young athletes. Arthroscopy: The Journal of Arthroscopic & Related Surgery. 2005;**21**(9):1066-1075

[82] Gudas R, Gudaitė A, Pocius A, et al. Ten-year follow-up of a prospective, randomized clinical study of mosaic osteochondral autologous transplantation versus microfracture for the treatment of osteochondral defects in the knee joint of athletes. American Journal of Sports Medicine. 2012;**40**(11):2499-2508

[83] Ulstein S, Årøen A, Røtterud JH, Løken S, Engebretsen L, Heir S. Microfracture technique versus osteochondral autologous transplantation mosaicplasty in patients with articular chondral lesions of the knee: A prospective randomized trial with long-term follow-up. Knee Surgery, Sports Traumatology, Arthroscopy. 2014;**22**(6):1207-1215

[84] Devitt BM, Bell SW, Webster KE, Feller JA, Whitehead TS. Surgical treatments of cartilage defects of the knee: Systematic review of randomised controlled trials. The Knee. 2017;**3**:508-517

[85] Mithoefer K, Williams RJ 3rd, Warren RF, Wickiewicz TL, Marx RG. High-impact athletics after knee articular cartilage repair: A prospective evaluation of the microfracture technique. American Journal of Sports Medicine. 2006;**34**(9):1413-1418

[86] Matsusue Y, Yamamuro T, Hama H. Arthroscopic multiple osteochondral transplantation to the chondral defect in the knee associated with anterior cruciate ligament disruption. Arthroscopy: The Journal of Arthroscopic & Related Surgery. 1993;**9**(3):318-321

[87] Hangody L, Fules P. Autologous osteochondral mosaicplasty for the treatment of full-thickness defects of weight-bearing joints: Ten years of experimental and clinical experience. Journal of Bone and Joint Surgery. America. 2003;**85-A**(2):25-32

[88] Richter DL, Tanksley JA, Miller MD. Osteochondral autograft transplantation: A review of the surgical technique and outcomes. Sports Medicine and Arthroscopy Review. 2016;**24**(2):74-78

[89] Versier G, Dubrana F. Treatment of knee cartilage defect in 2010. Orthopaedics & Traumatology: Surgery & Research. 2011;**97**(8):S140-S153

[90] Baltzer A, Ostapczuk M, Terheiden H, Merk H. Good short-to medium-term results after osteochondral autograft transplantation (OAT) in middle-aged patients with focal, non-traumatic osteochondral lesions of the knee. Orthopaedics & Traumatology: Surgery & Research. 2016;**102**(7):879-884

[91] Cognault J, Seurat O, Chaussard C, Ionescu S, Saragaglia D. Return to sports after autogenous osteochondral mosaicplasty of the femoral condyles: 25 cases at a mean follow-up of 9 years. Orthopaedics & Traumatology: Surgery & Research. 2015;**101**(3):313-317

[92] Pareek A, Reardon PJ, Maak TG, Levy BA, Stuart MJ, Krych AJ. Long-term outcomes after osteochondral autograft transfer: A systematic review at mean follow-up of 10.2 years. Arthroscopy: The Journal of Arthroscopic & Related Surgery. 2016;**32**(6):1174-1184

[93] Thaunat M, Couchon S, Lunn J, Charrois O, Fallet L, Beaufils P. Cartilage thickness matching of selected donor and recipient sites for osteochondral autografting of the medial femoral condyle. Knee Surgery, Sports Traumatology, Arthroscopy. 2007;**15**(4):381-386

[94] Hangody L, Kish G, Kárpáti Z, Udvarhelyi I, Szigeti I, Bély M. Mosaicplasty for the treatment of articular cartilage defects: Application in clinical practice. Orthopedics. 1998;**21**(7):751-756

[95] Marcacci M, Kon E, Zaffagnini S, et al. Multiple osteochondral arthroscopic grafting (mosaicplasty) for cartilage defects of the knee: Prospective study results at 2-year follow-up. Arthroscopy: The Journal of Arthroscopic & Related Surgery. 2005;**21**(4):462-470

[96] Berlet GC, Mascia A, Miniaci A. Treatment of unstable osteochondritis dissecans lesions of the knee using autogenous osteochondral grafts (mosaicplasty). Arthroscopy: The Journal of Arthroscopic & Related Surgery. 1999;**15**(3):312-316

[97] Miniaci A, Tytherleigh-Strong G. Fixation of unstable osteochondritis dissecans lesions of the knee using arthroscopic autogenous osteochondral grafting (mosaicplasty). Arthroscopy: The Journal of Arthroscopic & Related Surgery. 2007;**23**(8):845-851

[98] Solheim E, Hegna J, Øyen J, Austgulen OK, Harlem T, Strand T. Osteochondral autografting (mosaicplasty) in articular cartilage defects in the knee: Results at 5 to 9 years. The Knee. 2010;**17**(1):84-87

[99] Hangody L, Dobos J, Balo E, Panics G, Hangody LR, Berkes I. Clinical experiences with autologous osteochondral mosaicplasty in an athletic population: A 17-year prospective multicenter study. American Journal of Sports Medicine. 2010;**38**(6):1125-1133

[100] Kish G, Módis L, Hangody L. Osteochondral mosaicplasty for the treatment of focal chondral and osteochondral lesions of the knee and talus in the athlete: Rationale, indications, techniques, and results. Clinics in Sports Medicine. 1999;**18**(1):45-66

[101] Sadr KN, Pulido PA, McCauley JC, Bugbee WD. Osteochondral allograft transplantation in patients with osteochondritis dissecans of the knee. American Journal of Sports Medicine. 2016;**44**(11):2870-2875

[102] Gortz S, Bugbee WD. Allografts in articular cartilage repair. Journal of Bone and Joint Surgery. America. 2006;**88**(6):1374-1384

[103] Pallante AL, Bae WC, Chen AC, Gortz S, Bugbee WD, Sah RL. Chondrocyte viability is higher after prolonged storage at 37 degrees C than at 4 degrees C for osteochondral grafts. American Journal of Sports Medicine. 2009;**37**(1):24S-32S

[104] Garrity JT, Stoker AM, Sims HJ, Cook JL. Improved osteochondral allograft preservation using serum-free media at body temperature. American Journal of Sports Medicine. 2012;**40**(11):2542-2548

[105] LaPrade RF, Botker J, Herzog M, Agel J. Refrigerated osteoarticular allografts to treat articular cartilage defects of the femoral condyles. A prospective outcomes study. Journal of Bone and Joint Surgery. America. 2009;**91**(4):805-811

[106] Williams RJ 3rd, Dreese JC, Chen CT. Chondrocyte survival and material properties of hypothermically stored cartilage: An evaluation of tissue used for osteochondral allograft transplantation. American Journal of Sports Medicine. 2004;**32**(1):132-139

[107] Ball ST, Amiel D, Williams SK, et al. The effects of storage on fresh human osteochondral allografts. Clinical Orthopaedics and Related Research®. 2004;**418**:246-252

[108] Pallante AL, Chen AC, Ball ST, et al. The in vivo performance of osteochondral allografts in the goat is diminished with extended storage and decreased cartilage cellularity. American Journal of Sports Medicine. 2012;**40**(8):1814-1823

[109] Nielsen ES, McCauley JC, Pulido PA, Bugbee WD. Return to sport and recreational activity after osteochondral allograft transplantation in the knee. American Journal of Sports Medicine. 2017:**45**(7):1608-1614

[110] Emmerson BC, Gortz S, Jamali AA, Chung C, Amiel D, Bugbee WD. Fresh osteochondral allografting in the treatment of osteochondritis dissecans of the femoral condyle. American Journal of Sports Medicine. 2007;**35**(6):907-914

[111] Garrett JC. Fresh osteochondral allografts for treatment of articular defects in osteochondritis dissecans of the lateral femoral condyle in adults. Clinical Orthopaedics and Related Research®. 1994;**303**:33-37

[112] Frank RM, Lee S, Levy D, et al. Osteochondral allograft transplantation of the knee: Analysis of failures at 5 years. American Journal of Sports Medicine. Mar 2016;**45**(4): 864-874

[113] Peterson L. Technique of autologous chondrocyte transplantation. Techniques in Knee Surgery. 2002;**1**(1):2-12

[114] Mithöfer K, Minas T, Peterson L, Yeon H, Micheli LJ. Functional outcome of knee articular cartilage repair in adolescent athletes. American Journal of Sports Medicine. 2005;**33**(8):1147-1153

[115] Cole BJ, DeBerardino T, Brewster R, et al. Outcomes of autologous chondrocyte implantation in study of the treatment of articular repair (STAR) patients with osteochondritis dissecans. American Journal of Sports Medicine. 2012;**40**(9):2015-2022

[116] Peterson L, Brittberg M, Kiviranta I, Åkerlund EL, Lindahl A. Autologous chondrocyte transplantation biomechanics and long-term durability. American Journal of Sports Medicine. 2002;**30**(1):2-12

[117] Peterson L, Minas T, Brittberg M, Lindahl A. Treatment of osteochondritis dissecans of the knee with autologous chondrocyte transplantation. Journal of Bone and Joint Surgery. America. 2003;**85**(suppl 2):17-24

[118] Behery OA, Harris JD, Karnes JM, Siston RA, Flanigan DC. Factors influencing the outcome of autologous chondrocyte implantation: A systematic review. Journal of Knee Surgery. 2013;**26**(03):203-212

[119] Harris JD, Siston RA, Pan X, Flanigan DC. Autologous chondrocyte implantation: A systematic review. Journal of Bone and Joint Surgery. America. 2010;**92**(12):2220-2233

[120] Krishnan SP, Skinner JA, Bartlett W, et al. Who is the ideal candidate for autologous chondrocyte implantation? Journal of Bone and Joint Surgery. British. 2006;**88**(1):61-64

[121] Polousky JD, Albright J. Salvage techniques in osteochondritis dissecans. Clinics in Sports Medicine. 2014;**33**(2):321-333

[122] Chambers HG, Shea KG, Carey JL. AAOS clinical practice guideline: Diagnosis and treatment of osteochondritis dissecans. Journal of the American Academy of Orthopaedic Surgeons. 2011;**19**(5):307-309

[123] Nepple JJ, Milewski MD, Shea KG. Research in osteochondritis dissecans of the knee: 2016 update. The Journal of Knee Surgery. 2016;**29**(07):533-538

Chapter 8

Autologous Chondrocyte Implantation: Scaffold-Based Solutions

David C. Flanigan, Joshua S. Everhart and
Nicholas A. Early

Additional information is available at the end of the chapter

http://dx.doi.org/10.5772/intechopen.70276

Abstract

Autologous chondrocyte implantation is a surgical technique utilized for repair of articular cartilage defects. The originally described technique for autologous chondrocyte implantation involves applying a liquid suspension of the cultured chondrocytes to a cartilage defect and sealing the defect with a periosteum or collagen patch. Scaffolds for housing chondrocytes were introduced to allow for increased ease of delivery and application, to avoid leakage of chondrocytes out of the defect, and to allow for an implant that more closely mimics the non-uniform tissue architecture of healthy articular cartilage. In this chapter we describe the design, clinical outcomes, and commercial availability of various scaffolds reported in the clinical literature for autologous chondrocyte implantation.

Keywords: scaffold, MACI, MACT, autologous chondrocyte implantation, 3rd generation ACI

1. Introduction

Autologous chondrocyte implantation (ACI) is a two-stage articular cartilage repair technique for treatment of articular cartilage defects. Originally described by Brittberg et al. [1], it involves an initial surgery to harvest chondrocytes from a non-weight bearing portion of the distal femur, typically the intercondylar notch or medial or lateral margin of the trochlea. The cartilage extracellular matrix is then enzymatically digested within the laboratory to isolate the chondrocytes. The harvested chondrocytes are then cultured in a laboratory. In the second stage, a liquid suspension of chondrocytes is applied to the cartilage defect and is sealed in place with a soft tissue membrane cover [1]. Originally periosteum was utilized as the cover, though a collagen membrane was later introduced to minimize periosteal donor site morbidity and risk of periosteal patch hypertrophy [2]. Disadvantages of ACI with periosteum or collagen membrane covers with the use of a liquid cultured chondrocyte suspension include a high degree of technical difficulty, potential for leakage of chondrocytes, and non-uniform distribution of chondrocytes.

Scaffolds for housing chondrocytes were introduced for increased ease of delivery and application, to avoid leakage of chondrocytes out of the defect, and to allow homogeneous distribution of chondrocytes within the defect [3]. Additionally, there is some evidence that chondrocytes grown in monolayer culture do not fully regain their original phenotype [3, 4], which has prompted research in culture directly within a scaffold and design of implants that more closely mimics the non-uniform tissue architecture of healthy articular cartilage [3]. Use of a 3-dimensional structure for chondrocyte culture has been shown to maintain the chondrocyte differentiated phenotype [5]. Use of a scaffold is termed 'matrix-assisted autologous chondrocyte transplantation,' or the MACT procedure, and has been employed in clinical practice in Europe since 1998. The MACT procedure involves implantation of a chondrocyte seeded biocompatible scaffold in the articular defect [2]. The implant is fixed in place with fibrin glue with no membrane cover and allows for implantation with use of a mini-arthrotomy or arthroscopic implantation. The field of scaffold-based ACI has greatly expanded in recent years, with more than a dozen implants developed (**Table 1**). A wide variety of natural and synthetic materials have been utilized in MACT scaffolds; though clinical outcomes studies are generally favorable regardless of scaffold design, the number or published studies and length of follow-up vary widely among implants.

In this chapter, the design rationale, commercial availability, and clinical results of various scaffolds for use in MACT will be described. Of note, all implants described in this chapter follow a two-step implantation protocol (initial cartilage harvest and culturing of chondrocytes followed by a delayed implantation several weeks later). The single-stage implantation techniques with published outcomes data are either no longer commercially available (the CAIS implant) [6], or have yet to be marketed [7].

Scaffold content	Commercial name	Implantation steps
Porcine collagen I/III membrane	MACI	Two-steps
Three-dimensional collagen I based scaffold	NeoCart	Two-steps
Three-dimensional collagen I based scaffold	CaReS	Two-steps
Three-dimensional collagen I based scaffold	Novocart 3D	Two-steps
Hyaluronic acid based scaffold	Hyalograft C	Two-steps
Human fibrin and recombinant hyaluronic acid-based scaffold	BioCart II	Two-steps
Fibrin based gel	Chondron	Two-steps
Hydrogel of agarose and alginate	Cartipatch	Two-steps
Atelocollagen gel	Koken Atelocollagen Implant	Two-steps
Fibrin, polyglycolic/polylactic acid, polydioxanone	BioSeed-C	Two-steps

Table 1. Summary of MACT scaffolds.

2. Scaffolds utilized for autologous chondrocyte implantation

2.1. Porcine collagen I/III membranes

2.1.1. MACI

As of December 2016, matrix-assisted chondrocyte implantation (MACI; Vericel, Cambridge, MA) is currently the only FDA approved MACT technique for use in the United States. In this technique, chondrocytes are cultured ex-vivo in a monolayer and then seeded on one side of a porcine collagen I/III membrane (**Table 2**). At the second stage operation (re-implantation), the side seeded with chondrocytes (the roughened side) is placed against the subchondral bone surface and the graft is secured with fibrin glue [8]. The implantation may be performed arthroscopically or with a mini-arthrotomy, and recent work demonstrates MACI grafts may be safely applied with use of carbon dioxide insufflation arthroscopy [9]. Regardless of technique, gentle handling of the graft is recommended, as excessive or forceful handling of the graft causes a significant decrease in viable chondrocytes [10]. A histologic study of 56 MACI patients up to 6 months after surgery demonstrated that chondrocytes appeared well-integrated and maintained chondrocyte phenotype [11]. Hyaline-like cartilage production began as early as 21 days after implantation, and there was 75% hyaline-like cartilage regeneration at 6 months [11]. Another histologic study of 33 second-look biopsies at median 15 months after surgery found a median ICRS histological grade of 57 which did not correlate with an arthroscopic ICRS grade of normal in 30% of cases and nearly normal in 51% of cases [12].

Several comparative studies have been performed with MACI, all of which demonstrated encouraging results (**Table 3**). However, it should be noted that use of MACI in clinical practice tends to be in larger defects (mean 5.64 cm^2) than lesions treated in clinical trials (weighted mean 4.89 cm^2) [13]. Approval by the FDA was based primarily on results of the SUMMIT trial, reported by Saris et al. [14]. In this randomized trial, 144 patients with high grade femoral condylar defects were randomized to MACI or microfracture and followed for 2 years; mean defect size was equivalent between groups (4.9 cm^2 MACI vs. 4.7 cm^2 microfracture) [14]. At final follow-up there was significantly better improvement in KOOS symptom scores with MACI, lower failure rates, yet no difference in repair quality as assessed by histology or MRI versus microfracture [14]. A randomized controlled trial was performed by Bartlett et al. with comparison of ACI-C (ACI with collagen cover) and MACI for treatment

Commercial name	Manufacturer	Structure	Expansion	Availability
MACI	Vericel, Cambridge, MA (Formerly provided by Verigen Transplantation Service, Copenhagen, Denmark)	Porcine-derived collagen I/III bilayer	Cells are expanded in monolayer then seeded onto porous side of collagen membrane	FDA approved for use in the USA. Available in Europe and Australia

FDA, U.S. Food and Drug Administration.

Table 2. MACT with porcine collagen I/III membrane scaffold.

Level of evidence	Author	Implant and sample size	Mean follow-up	Outcome
1	Ebert et al. [20]	19 MACI-standard WB; 18 MACI-accelerated WB	2 years	Randomized trial of standard 8 week return to weight bearing versus accelerated 6 week return to weight bearing. No difference in symptom improvement.
1	Wondrasch et al. [21]	15 MACI-standard WB; 16 MACI-accelerate WB	5 years	Randomized trial of 6 versus 10 week return to weight bearing. No difference in symptom improvement between groups. MOCART score decreased from years 2 to 5 which did not correlate with symptom scores
1	Akgun et al. [18]	7 MACI; seven mesenchymal stem cell	2 years	Small randomized trial of MACI versus stem cells (also seeded onto a collagen scaffold. Stem cell group had greater symptom improvement at 6 months but similar improvement at final follow-up.
1	Basad et al. [15]	40 MACI; 20 microfracture	2 years	At 24 months, greater improvements seen with MACI in Tegner activity score, subjective symptoms scores and ICRS scores on 2nd look arthroscopy.
1	Saris et al. [14]	72 MACI; 72 microfracture	2 years	Greater improvement in KOOS scores, lower failure rate with MACI (12.5%) versus microfracture (31.9%). Similar MRI and histologic outcomes.

Table 3. Outcomes of MACT with collagen I/III membrane scaffold (MACI) from level 1 prospective clinical studies.

of high grade chondral defects. Mean defect size was 6.0 cm^2 for the MACI group and 6.1 cm^2 for the ACI-C group [8]. At 1 year follow-up both groups demonstrated significant improvement in Cincinnati knee scores and similar re-operation rates (9% for both groups) [8]. Basad et al. performed a randomized study of MACI versus microfracture with 2 years follow-up on high grade defects 4–10 cm^2 [15]. The MACI group in this study had greater improvements in symptom scores, activity scores, and ICRS surgeon grading of cartilage appearance at second look arthroscopy [15]. In a comparative imaging and clinical study of MACI versus osteochondral autograft transfer (OAT) by Salzmann et al., superior Lysholm symptoms scores were observed in the MACI group; patients in this study were matched for demographics, but MACI-treated lesions were >3 cm^2 and OAT-treated lesions were <3 cm^2 [16]. For treatment of chondromalacia patella, Macmull et al. noted a higher rate of good-excellent patient symptom scores with MACI (56.5%) than ACI-C (40%). Higher rates of clinical failure (poor patient-rated symptoms) were noted with lateral facet lesions, and the authors did not report distribution of lesions (medial facet, lateral facet, or multiple facets) by treatment group [17]. Finally, Akgun et al. report a small randomized trial of MACI versus autologous mesenchymal stem cells (also seeded onto a collagen scaffold) with 2 years follow-up [18]. The stem cell group had greater symptom improvement at 6 months but similar improvement at final follow-up; no clinical failures were noted in either group [18].

Several randomized trials of delayed versus accelerated weight-bearing after MACI have been performed (**Table 3**). A randomized trial of 6 week versus 8 week return to full weight bearing found no significant difference in failure rates or symptom improvement at 2 years (interim 12-month results reported in an earlier publication [19]); the study authors concluded accelerated weight bearing after MACI is safe [20]. Another trial of 6 week versus 10 week return to full weight bearing with 5 years follow-up after MACI similarly found no difference in symptom improvement between groups [21]. The authors note that MRI-based MOCART scores decreased from years 2 to 5 but did not correlate with symptom scores [21].

Several case series have reported also reported good results with MACI (**Table 3**). The series with the longest follow-up is reported by Gille et al.; of 19 cases with mean 16 years follow-up, 21% underwent knee arthroplasty (4/19), with durable symptom improvement in the remaining 15 patients [22]. In another series of MACI patients, Basad et al. report durable improvements in activity and symptoms scores and a failure rate of 18.5% at 5 years with MACI [23]. Behrens et al. similarly report 8/11 patients rated their current knee function as 'much better or better' than their pre-operative function at 5 years follow-up [24]. A larger case series by Ebert et al. of 41 patients and 5 years follow up (35/41, 85% with 5 years follow-up) reported significant improvements in knee function, a 12% rate of graft hypertrophy at 5 years, and a graft failure rate of 3% at 5 years [25]. Durable results are seen with arthroscopic implantation of MACI scaffolds, as Ebert et al. report stable clinical improvement at 5 years follow-up and a failure rate of 6.4% [26]. Ventura et al. note improvement in Lysholm symptom scores at 2 years but no change in Tegner activity scores in a series of 53 patients; a high rates of subchondral abnormalities were noted on MRI at 1 year (70% of cases) which did not correlate with clinical symptoms [27].

For the patellofemoral joint, Meyerkort et al. report durable improvement in symptoms at 5 years with MACI; clinical improvement did not correlate with MRI assessment of graft appearance at 5 years [28]. Gigante et al. published results of treatment of patellar defects with MACI and concomitant distal realignment; at 3 years, there was significant improvement in symptoms in most patients and one clinical failure (7%) [29].

As a salvage operation in young patients with medial compartment osteoarthritis, Bauer et al. report significant clinical improvement at 5 years with combined high tibial osteotomy and MACI; however, they note declining results and high graft failure over time for this salvage operation [30]. Finally, outcomes for MACI and concomitant bone grafting for treatment of osteochondral lesions with use of a bilayer 'sandwich' technique have also been reported. Vijayan et al. report outcomes with use of two MACI membranes and impaction bone grafting of osteochondral lesions greater than 8 mm depth; at a mean 5.2 year's follow-up, 12/14 patients had good to excellent results with one graft failure [31].

2.2. Three-dimensional collagen I based scaffolds

2.2.1. NeoCart

NeoCart (Histogenics Corporation, Waltham, Massachusetts) is an MACT implant that consists of a three-dimensional bovine collagen I scaffold (**Table 4**). Rather than being cultured

Commercial name	Manufacturer	Structure	Expansion	Availability
NeoCart	Histogenics Corporation, Waltham, Massachusetts	Bovine collagen type I matrix	Cells are expanded directly on 3D scaffold via a custom bioreactor	Ongoing phase III clinical trials; not yet approved by the FDA
CaReS	Arthro Kinetics (Ars Arthro, Esslingen, Germany)	Rat collagen type I matrix	Cells are mixed with collagen which forms a gel and cultured for 2 weeks	SFDA certified; not yet approved by the FDA
Novocart 3D	B. Braun-Tetec, Reutlingen, Germany	Collagen-chondroitin sulfate scaffold	Initial monolayer culture followed by seeding onto scaffold; re-implantation 3–4 weeks after harvest	Available in Europe, ongoing phase III clinical trials.

FDA, U.S. Food and Drug Administration; SFDA, State Food and Drug Administration of China; 3D, three-dimensional.

Table 4. MACT with three-dimensional collagen 1 scaffold.

in a monolayer, the scaffold is seeded initially with chondrocytes which then proliferate in a custom bioreactor [32]. The bioreactor is designed to incubate the scaffold in a low-oxygen tension environment with varying pressure to mimic the native intra-articular environment with the goal of preserving the chondrocyte phenotype [33]. At the time of implantation, the graft is fixed to the defect with a proprietary adhesive (CT3 bioadhesive, Histogenics). A randomized phase II trial by Crawford et al. of distal femoral lesions treated with NeoCart versus microfracture demonstrated superior improvement in IKDC and KOOS scores at 24 months with NeoCart and no difference in adverse events between groups (**Table 5**) [33]. A small case series (8 patients) with 2 years follow-up demonstrated significant symptom improvement from baseline and no cases of graft hypertrophy or arthrofibrosis (**Table 5**) [32]. Defect fill was noted to be moderate (33–66%) in 1/8 cases and poor (<33%) in 1/8 cases. A longitudinal

Level of evidence	Author	Implant and sample size	Mean follow-up	Outcome
1	Crawford et al. [33]	21 NeoCart; 9 microfracture	2 years	Randomized trial of distal femoral lesions. Greater IKDC and KOOS improvement at 2 years with NeoCart.
3	Flohe et al. [35]	9 CaReS; 11 MACI	1 year	No difference in clinical outcomes between groups.
3	Petri et al. [36]	17CaReS; 10 microfracture	3 years	Comparative trial for patellofemoral defects. No difference in groups between IKDC, SF-36, or Cincinnati knee scores at 3 years follow-up.

Table 5. MACT clinical outcome studies with three-dimensional collagen 1 scaffold.

clinical and MRI-based study with 5 years follow-up by Anderson et al. demonstrate that clinical improvement and graft appearance on MRI both evolve over the first 24 months after surgery [34]. Both clinical scores and MRI appearance appeared stable from 24 to 60 months follow-up [34].

2.2.2. *CaReS*

The Cartilage Regeneration System (CaReS, Ars Arthro, Esslingen, Germany) utilizes a rat-derived collagen I gel rather than the bovine collagen matrix utilized by NeoCart (**Table 4**). The harvested chondrocytes are similarly seeded into the collagen gel and cultured in this 3-dimensional environment with the intention of preserving cartilage phenotype. In a small comparative study of CaReS (9 patients) versus MACI (11 patients) with 1 year follow-up, Flohe et al. demonstrate significant improvement in symptoms with no difference between groups (**Table 5**) [35]. A small comparative study of microfracture (n = 10) vs. CaReS (n = 17) for patellofemoral lesions found significant improvements in symptoms from baseline with no difference in outcomes between groups [36]. In a multicenter clinical trial, Schneider et al. report outcomes of 116 at mean 30.6 month follow-up from 9 different centers; mean defect size in the trial was 5.4 cm^2 [37]. At final follow-up there was significant improvement in IKDC, VAS and SF-36 scores and a patient satisfaction rate of 80%. A total of 8 revision arthroscopies were performed for pain with 2 cases of implant hypertrophy and 2 cases of early failure [37]. In an imaging based outcome study, Welsch et al. compared 3T MRI results at 2 years for Hyalograft C versus CaReS and found greater T2 relaxation times for CaReS despite similar clinical outcomes between groups [38].

2.2.3. *Novocart 3D*

The Novocart 3D implant (B. Braun-Tetec, Reutlingen, Germany) is a collagen-chondroitin sulfate sponge (**Table 4**). After chondrocyte harvest, cells are initially cultured in a monolayer and then seeded onto the collagen-chondroitin sulfate scaffold at a density of $0.5–3.0 \times 10^6$ cells/cm^2, after which the scaffold is cultivated in serum for 2 days before shipment for re-implantation [39]. Niethammer et al. performed several MRI-based studies of graft maturation and graft filling with Novocart 3D. In a 3 years prospective MRI study, graft maturation as assessed by T2 mapping required at least 1 year [40]. In a 2 years prospective MRI study, incomplete graft filling as assessed by MRI was common (55.7%) at 2 years and did not correlate with clinical results; the authors noted that graft thickness appeared to increase throughout the 2 years follow-up period [41]. A 2 years follow-up MRI study showed a 25% graph hypertrophy rate in Novocart 3D patients (11/44 patients), with higher hypertrophy rates in cases of acute traumatic defects or osteochondritis dissecans [42].

In a small non-randomized comparative study, Panagopoulos et al. report outcomes of Novocart 3D (n = 9) and ACI-P (periosteal cover) (n = 11) and mean 37.5 months follow-up (**Table 5**) [43]. No significant difference in Tegner, Lysholm, or IKDC scores was noted between groups. The patient population consisted of high demand athletes and soldiers, with low rates of return to pre-injury activity levels (6/19, 31.5%) [43]. In a comparative study of 40 pediatric (<20 years old) patients treated with Novocart 3D versus 40 matched adult historical controls who also underwent Novocart for similar size/location lesions, both groups had significant improvement in VAS and IKDC scores at 36 months, but the pediatric group had greater improvement than the adult group at final follow-up [44]. A case series of 23 patients with 2 years follow-up by Zak et al. report improvement in symptoms scores as well as activity scores versus baseline with use of Novocart 3D [39]. At final follow-up, hypertrophy was noted via MRI in 16% and incomplete filling (>50%) in 20% of patients [39]. A large case series by Angele et al. of 433 patients with mean 6.9 months follow-up (max 2.5 years) found an 8.5% re-operation rate, a 6% graft failure rate in patients with >12 months follow-up [45]. Finally, in a case series with 2 years follow-up, Niethammer noted that clinical outcomes at 2 years were worse for patients who returned to sport/physical activities at earlier than 12 months after surgery [46].

2.3. Hyaluronic acid or fibrin based scaffolds

2.3.1. Hyalograft C

The Hyalograft C scaffold is based on the benzylic ester of hyaluronic acid (HYAFF 11; Fidia Advanced Biopolymers Laboratories, Padova, Italy) (**Table 6**). The resulting scaffold is a meshwork of 20 micrometer diameter fibers. The cells are cultured directly on the scaffold with resulting collagen II and aggrecan production [5]. The implant is naturally adhesive and does not require an additional adhesive at time of implantation. Clinical outcomes of Hyalograft C were encouraging, with superior results in comparison to microfracture [47] and comparable results to MACI [48] or traditional ACI with a periosteum cover (**Table 7**) [49]. However, production of this implant has been discontinued by the manufacturer in favor of further development of a single-stage delivery system (no published clinical outcomes data available for the single-stage system).

Commercial name	Manufacturer	Structure	Expansion	Availability
Hyalograft C	Anika Therapeutics (Fidia Advanced Biopolymers Laboratories, Padova, Italy)	Benzylic ester of hyaluronic acid (HYAFF) combined with expanded patient cells	Cells seeded and cultured directly on scaffold	No longer commercially available; production discontinued
BioCart II	Histogenics Corporation, Waltham, MA (merger with former supplier, ProChon Biotech)	Human fibrin and recombinant hyaluronic acid-based scaffold	Cells cultured in human serum and growth factor FGF2v1, then seeded onto scaffold	Available in Italy, Greece, and Israel; ongoing clinical trials in the United States; not yet approved by the FDA
Chondron	Sewon Cellontech, Seoul, Korea	Fibrin based gel	Cells cultured in serum; at time of surgery, suspension is mixed 1:1 with fibrin	Available in Korea

AIFA, Italian Medicines Agency; FDA, U.S. Food and Drug Administration.

Table 6. Hyaluronic acid or fibrin-based scaffolds.

Level of evidence	Author	Implant and sample size	Mean follow-up	Outcome
2	Kon et al. [47]	21 Hyalograft C; 20 microfracture	7.5 years	Return to sport was a median 8 months for microfracture, 12.5 months for Hyalograft C. Symptom improvement with microfracture deteriorated with time whereas Hyalograft C was durable.
3	Kon et al. [48]	22 Hyalograft C; 39 MACI	5 years	All patients 40 years or older, treated with mini-open MACI or arthroscopic Hyalograft C. Overall failure rate 20%, similar symptom improvement seen in both treatment groups.
3	Ferruzzi et al. [49]	50 Hyalograft C; 48 ACI-P	2–5 years	Similar IKDC improvement at 2+ years. Greater symptom improvement in first 12 months in Hyalograft C (arthroscopic) group versus ACI-P (mini-open)

Table 7. MACT clinical outcome studies with hyaluronic acid or fibrin-based scaffolds.

2.3.2. *BioCart II*

An implant called BioCart II (Histogenics Corporation, Waltham, MA formerly supplied by ProChon Biotech prior to merger with Histogenics) is comprised of a scaffold of recombinant hyaluronan with fibrin to form a sponge (**Table 6**). Cells are initially cultured in human serum with recombinant fibroblast growth factor 2 variant (FGF2v1) and then seeded onto the scaffold prior to implantation with a mini-open approach. A small 1 year outcome study by Nehrer et al. of 8 patients demonstrated significant improvement in IKDC and Lysholm scores; 3 patients had a transient effusion post-operatively and there were no clinical failures (**Table 7**) [50]. A case series by Eshed et al. of patients who underwent MRI evaluation at mean 17.3 months after surgery (range 6–48 months) found continued maturation of cartilage with time (>1 year versus <1 year) and higher IKDC scores in patients with >12 months follow-up and without a history of prior cartilage surgeries [51].

2.3.3. *Chondron*

The Chondron scaffold is a fibrin-based gel (Sewon Cellontech Co. Ltd., Seoul, Korea) (**Table 6**). Chondrocytes are first cultured separately in a specialized serum (CRM kit, Sewon Cellontech, Korea). At the time of surgery the serum and cultured chondrocytes are mixed 1:1 with fibrin and injected directly onto the defect. In addition to typical preparation of the defect for ACI, several holes are drilled into the subchondral bone to improve adherence [52]. Choi et al. report a multicenter study of 98 patients with mean 24 month follow-up treated with Chondron (**Table 7**) [52]. Symptom improvement increased with time, with greater improvement noted with >25 months follow-up versus <25 months. Complication rates were low with one early repeat operation (1%) and two cases of symptomatic catching (2%) [52]. Similar findings were reported in a series by Kim et al., with no graft-related complications among 30 patients at 24 months follow up; a second look arthroscopy at 12 months showed nearly normal cartilage in 8/10 patients [53]. A small series by Konst et al. of 9 patients with osteochondral defects (mean depth 0.9 cm) treated with autologous bone grafting as well as Chondron showed satisfactory short term results at 12 months; there was one treatment failure which was converted to a unicompartmental knee arthroplasty [54].

2.4. Alginate based scaffolds

2.4.1. Cartipatch

Cartipatch (TBF Tissue Engineering, Mions, France) is a MACT implant with a scaffold composed of agarose and alginate (**Table 8**). Chondrocytes are first cultured in a monolayer and then mixed with a hydrogel of agarose and alginate. The hydrogel can be manipulated at 37°C and will solidify around 25°C, allowing formation of complex/irregular shapes with the scaffold. A multicenter randomized trial with 2 years follow-up was recently published by Clave et al. (**Table 9**) [55]. In this study, 30 patients were randomized to Cartipatch and 25 to mosaicplasty; all patients had isolated high grade femoral condylar defects 2.5–7.5 cm^2 in size. At 2 years, there was significantly greater improvement in IKDC scores with mosaicplasty than Cartipatch, though both groups had significant improvement over baseline. A total of 12 adverse events were reported for the Cartipatch groups and six in the mosaicplasty group [55]. An earlier case series by Selmi et al. reported 2 years outcomes of 17 patients treated with Cartipatch with a mean defect size of 3 cm^2 [56]. All patients had significant symptom improvement with no clinical failures; second look biopsies in 13 patients had mostly hyaline-like cartilage in 62% of cases (8/13) [56].

Commercial name	Manufacturer	Structure	Expansion	Availability
Cartipatch	Tissue Bank of France (TBF) Tissue Engineering, Mions, France	Alginate-agarose hydrogel combined with autologous cells	Two-step procedure; reduces cell leakage and implantation time	Ongoing phase III clinical trials; not yet approved by the FDA

FDA, U.S. Food and Drug Administration.

Table 8. Alginate hydrogel.

Level of evidence	Author	Implant and sample size	Mean follow-up	Outcome
1	Clave et al. [55]	30 Cartipatch; 25 mosaicplasty	2 years	Both groups showed improvement in IKDC scores over baseline though mosaicplasty had greater symptom improvement than Cartipatch at 2 years for femoral lesions 2.5–7.5 cm^2.
4	Selmi et al. [56]	17 Cartipatch	2 years	Multicenter study. Significant symptom improvement in all patients, no clinical failures. Second look biopsies showed mostly hyaline-like cartilage in 8/13 patients (62%).

Table 9. MACT clinical outcome studies with alginate-based scaffolds.

2.5. Atelocollagen gel

2.5.1. *Koken Atelocollagen Implant*

The MACT technique with use of the Koken Atelocollagen Implant (Koken, Tokyo, Japan) is similar to the ACI-P (periosteum cover) technique, but chondrocytes are suspended in atelocollagen gel rather than a liquid to obtain uniform distribution of chondrocytes within the defect and theoretically reduce risk of leakage (**Table 10**). In this technique, after initial isolation of chondrocytes from cartilage biopsy, the chondrocyte suspension is mixed 1:4 with a 3% bovine atelocollagen solution (Koken, Tokyo, Japan) [57]. Chondrocytes are expanded in this mixture for 28 days; the final product (the Koken Atelocollagen Implant) is an opaque implant with a jelly-like consistency. The Koken Atelocollagen Implant is implanted with a mini-arthrotomy and requires a periosteum cover to contain the atelocollagen-based scaffold within the defect [57]. A multicenter trial in Japan reported by Tohyama et al. reports use of the Koken Atelocollagen Implant and periosteum cover in 27 patients (**Table 11**) [57]. Overall there was a significant improvement in Lysholm scores at final 2 years follow-up. On second look arthroscopy, 24% of repair sites were ICRS grade normal and 48% were nearly normal. There was one case of graft hypertrophy, two cases of graft detachment, and two cases of abnormal or severely normal ICRS grade on second look arthroscopy [57]. Recently, Tadenuma et al. report clinical and imaging outcomes of 8 patients (11 knees) at mean 5.9 years after surgery [58]. The authors note significant improvement in Lysholm scores over baseline with one clinical failure (9%) and one traumatic repeat injury 7 years after surgery (9%). The authors report a correlation between T1 values of the repair site on MRI and clinical outcomes but no correlation between T2 values and outcomes [58].

Commercial name	Manufacturer	Structure	Expansion	Availability
Koken Atelocollagen Implant	Koken, Tokyo, Japan	Atelocollagen gel (3% type 1 bovine collagen gel)	Chondrocyte suspension is initially mixed 1:4 with 3% atelocollagen solution. The mixture is cultured for 4 weeks and thickens to a jelly-like consistency over that time.	Available in Japan

Table 10. Atelocollagen based scaffold.

Level of evidence	Author	Implant and sample size	Mean follow-up	Outcome
4	Tohyama et al. [57]	27 Koken Atelocollagen Implant	2 years	Multicenter study. Symptom scores (Lysholm) improved at 2 years from baseline. Two cases of graft detachment (7.4%). Two remaining cases were graded abnormal or severely abnormal on second look arthroscopy (8%, 2/25).
4	Tadenuma et al. [58]	11 Koken Atelocollagen Implant	5.9 years	Improved Lysholm scores at final follow-up with one clinical failure (9%). T1 scores on MRI at final follow-up correlated with clinical scores but T2 scores did not.

Table 11. MACT clinical outcome studies with alginate-based scaffolds.

2.6. Polyglycolic/polylactic acid and polydioxanone based scaffold

2.6.1. BioSeed-C

The BioSeed-C (BioTissue Technologies GmbH, Freiburg, Germany) MACT scaffold is comprised polyglycolic/polylactic acid (polyglactin, vicryl), and polydioxanone (**Table 12**). Harvested chondrocytes are first expanded in serum and then seeded into the polymer scaffold with fixation by fibrin. The scaffold is available in a standard rectangular shape (2 cm × 3 cm × 0.2 cm thickness) can be implanted arthroscopically or with a mini-arthrotomy. The defect must be contoured to a rectangular shape (more than one scaffold can be used as needed for larger defects) and corners of the scaffold are secured with transosseous resorbable suture loops [59].

In a comparative non-randomized study of ACI-P versus BioSeed-C with minimum 2 years follow up, Erggelet et al. report similar improvement in symptom scores (**Table 13**) [60]. The graft failure rate was similar between groups (3/42 ACI-P; 2/40 BioSeed-C), but re-operation rates were twice as high in the ACI-P group, primarily due to graft hypertrophy [60]. A smaller randomized study of ACI-P (n = 10) versus BioSeed-C (n = 9) with 2 years follow-up by Zeifang et al. found similar improvement in symptoms between groups (per IKDC score) at both 1 and 2 years [61]. In contrast to the findings reported by Erggelet et al. [60], re-operation rates were higher in the BioSeed C group (3/11 patients) versus ACI-P (1/10 patients) [61].

Commercial name	Manufacturer	Structure	Expansion	Availability
BioSeed-C	BioTissue AG (BioTissue Technologies, GmbH, Freiburg, Germany)	Fibrin, polyglycolic/polylactic acid and polydioxanone-based material combined with culture-expanded autologous chondrocytes and suspended in fibrin.	Chondrocytes cultured in serum then subsequently seeded into scaffold.	CE mark approval; not yet approved by the FDA.

CE, Conformité Européenne; FDA, U.S. Food and Drug Administration.

Table 12. Scaffolds with polyglycolic/polylactic acid and polydioxanone.

Level of evidence	Author	Implant and sample size	Mean follow-up	Outcome
2	Zeifang et al. [61]	11 BioSeed-C; 9 ACI-P	2 years	Similar IKDC symptom improvement in both groups at 1 year and 2 years. Higher re-operation rate in BioSeed C group.
3	Erggelet et al. [60]	40 BioSeed-C; 42 ACI-P	36 m ACI-P 24 m BioSeed-C	Twice as many re-operations required for ACI-P versus BioSeed-C. Three graft failures in ACI-P group and two in BioSeed-C group. Equivalent improvement in symptom scores between groups.

Table 13. MACT clinical outcome studies with polyglycolic/polylactic acid and polydioxanone based scaffold.

Several case series have also been reported for BioSeed-C (**Table 13**). Ossendorf et al. report a case series of 40 patients treated with BioSeed-C with 2 years follow-up; symptom scores were significantly improved at both 1 and 2 years after baseline [59]. Reoperations occurred in 12.5% of patients including synovectomy (n = 2), debridement (n = 1), total knee arthroplasty (n = 1), and graft removal (n = 1) [59]. The mid-term outcomes of the same patient cohort with 4-years follow-up were reported by Kreuz et al. [62]. The authors note a durable symptom improvement over 4 years and a high rate of graft filling (mostly or completely filled in 43/44 patients on MRI assessment) [62]. In the subgroup analysis of 19 patients in this cohort with baseline osteoarthritis and a high grade focal defect, Kreuz et al. noted symptom improvement at 6–12 months which remained stable at 4 years as well as two clinical failures that went on to total knee arthroplasty (10.5%) [63].

3. Conclusions

In conclusion, short and mid-term clinical outcomes studies of MACT therapies for cartilage defects of the knee have been encouraging. However, commercial availability of MACT procedures is highly variable with respect to geographic region. Recent approval was granted in December 2016 by the FDA for use of MACI in the United States. To date this is the only MACT therapy available in this region. Availability is greater for multiple MACT therapies in Europe, though European Medicine Agency marketing approval for MACI was recently suspended in June 2016.

Author details

David C. Flanigan[1,2]*, Joshua S. Everhart[1] and Nicholas A. Early[2]

*Address all correspondence to: david.flanigan@osumc.edu

1 Sports Medicine, The Ohio State University Wexner Medical Center, Columbus, OH, United States

2 Department of Physical Medicine and Rehabilitation, Washington University, St. Louis, MO, United States

References

[1] Brittberg M, Lindahl A, Nilsson A, Ohlsson C, Isaksson O, Peterson L. Treatment of deep cartilage defects in the knee with autologous chondrocyte transplantation. The New England Journal of Medicine. 1994;**331**:889-895

[2] Ruta DJ, Villarreal AD, Richardson DR. Orthopedic surgical options for joint cartilage repair and restoration. Physical Medicine and Rehabilitation Clinics of North America. 2016;**27**:1019-1042

[3] Kon E, Filardo G, Di Martino A, Marcacci M. ACI and MACI. The Journal of Knee Surgery. 2012;**25**:17-22

[4] Kon E, Verdonk P, Condello V, et al. Matrix-assisted autologous chondrocyte transplantation for the repair of cartilage defects of the knee: Systematic clinical data review and study quality analysis. The American Journal of Sports Medicine. 2009;**37**(Suppl 1): 156S-166S

[5] Grigolo B, Lisignoli G, Piacentini A, et al. Evidence for redifferentiation of human chondrocytes grown on a hyaluronan-based biomaterial (HYAff 11): Molecular, immunohistochemical and ultrastructural analysis. Biomaterials. 2002;**23**:1187-1195

[6] Cole BJ, Farr J, Winalski CS, et al. Outcomes after a single-stage procedure for cell-based cartilage repair: A prospective clinical safety trial with 2-year follow-up. The American Journal of Sports Medicine. 2011;**39**:1170-1179

[7] Chiang H, Liao CJ, Hsieh CH, Shen CY, Huang YY, Jiang CC. Clinical feasibility of a novel biphasic osteochondral composite for matrix-associated autologous chondrocyte implantation. Osteoarthritis and Cartilage. 2013;**21**:589-598

[8] Bartlett W, Skinner JA, Gooding CR, et al. Autologous chondrocyte implantation versus matrix-induced autologous chondrocyte implantation for osteochondral defects of the knee: A prospective, randomised study. Journal of Bone and Joint Surgery. British Volume (London). 2005;**87**:640-645

[9] Vascellari A, Rebuzzi E, Schiavetti S, Coletti N. Implantation of matrix-induced autologous chondrocyte (MACI®) grafts using carbon dioxide insufflation arthroscopy. Knee Surgery, Sports Traumatology, Arthroscopy. 2014;**22**:219-225

[10] Hindle P, Hall AC, Biant LC. Viability of chondrocytes seeded onto a collagen I/III membrane for matrix-induced autologous chondrocyte implantation. Journal of Orthopaedic Research. 2014;**32**:1495-1502

[11] Zheng MH, Willers C, Kirilak L, et al. Matrix-induced autologous chondrocyte implantation (MACI): Biological and histological assessment. Tissue Engineering. 2007;**13**:737-746

[12] Enea D, Cecconi S, Busilacchi A, Manzotti S, Gesuita R, Gigante A. Matrix-induced autologous chondrocyte implantation (MACI) in the knee. Knee Surgery, Sports Traumatology, Arthroscopy. 2012;**20**:862-869

[13] Foldager CB, Farr J, Gomoll AH. Patients scheduled for chondrocyte implantation treatment with MACI have larger defects than those enrolled in clinical trials. Cartilage. 2016;**7**:140-148

[14] Saris D, Price A, Widuchowski W, et al. Matrix-applied characterized autologous cultured chondrocytes versus microfracture: Two-year follow-up of a prospective randomized trial. The American Journal of Sports Medicine. 2014;**42**:1384-1394

[15] Basad E, Ishaque B, Bachmann G, Stürz H, Steinmeyer J. Matrix-induced autologous chondrocyte implantation versus microfracture in the treatment of cartilage defects of the knee: A 2-year randomised study. Knee Surgery, Sports Traumatology, Arthroscopy. 2010;**18**:519-527

[16] Salzmann GM, Paul J, Bauer JS, et al. T2 assessment and clinical outcome following autologous matrix-assisted chondrocyte and osteochondral autograft transplantation. Osteoarthritis and Cartilage. 2009;**17**:1576-1582

[17] Macmull S, Jaiswal PK, Bentley G, Skinner JA, Carrington RW, Briggs TW. The role of autologous chondrocyte implantation in the treatment of symptomatic chondromalacia patellae. International Orthopaedics. 2012;**36**:1371-1377

[18] Akgun I, Unlu MC, Erdal OA, et al. Matrix-induced autologous mesenchymal stem cell implantation versus matrix-induced autologous chondrocyte implantation in the treatment of chondral defects of the knee: A 2-year randomized study. Archives of Orthopaedic and Trauma Surgery. 2015;**135**:251-263

[19] Edwards PK, Ackland TR, Ebert JR. Accelerated weightbearing rehabilitation after matrix-induced autologous chondrocyte implantation in the tibiofemoral joint: Early clinical and radiological outcomes. The American Journal of Sports Medicine. 2013;**41**:2314-2324

[20] Ebert JR, Edwards PK, Fallon M, Ackland TR, Janes GC, Wood DJ. Two-year outcomes of a randomized trial investigating a 6-week return to full weightbearing after matrix-induced autologous chondrocyte implantation. The American Journal of Sports Medicine. 2017;**45**:838-848

[21] Wondrasch B, Risberg MA, Zak L, Marlovits S, Aldrian S. Effect of accelerated weight-bearing after matrix-associated autologous chondrocyte implantation on the femoral condyle: A prospective, randomized controlled study presenting MRI-based and clinical outcomes after 5 years. The American Journal of Sports Medicine. 2015;**43**:146-153

[22] Gille J, Behrens P, Schulz AP, Oheim R, Kienast B. Matrix-associated autologous chondrocyte implantation: A clinical follow-up at 15 years. Cartilage. 2016;**7**:309-315

[23] Basad E, Wissing FR, Fehrenbach P, Rickert M, Steinmeyer J, Ishaque B. Matrix-induced autologous chondrocyte implantation (MACI) in the knee: Clinical outcomes and challenges. Knee Surgery, Sports Traumatology, Arthroscopy. 2015;**23**:3729-3735

[24] Behrens P, Bitter T, Kurz B, Russlies M. Matrix-associated autologous chondrocyte transplantation/implantation (MACT/MACI)—5-year follow-up. The Knee. 2006;**13**:194-202

[25] Ebert JR, Robertson WB, Woodhouse J, et al. Clinical and magnetic resonance imaging-based outcomes to 5 years after matrix-induced autologous chondrocyte implantation to address articular cartilage defects in the knee. The American Journal of Sports Medicine. 2011;**39**:753-763

[26] Ebert JR, Fallon M, Wood DJ, Janes GC. A prospective clinical and radiological evaluation at 5 years after arthroscopic matrix-induced autologous chondrocyte implantation. The American Journal of Sports Medicine. 2017;**45**:59-69

[27] Ventura A, Memeo A, Borgo E, Terzaghi C, Legnani C, Albisetti W. Repair of osteochondral lesions in the knee by chondrocyte implantation using the MACI® technique. Knee Surgery, Sports Traumatology, Arthroscopy. 2012;**20**:121-126

[28] Meyerkort D, Ebert JR, Ackland TR, et al. Matrix-induced autologous chondrocyte implantation (MACI) for chondral defects in the patellofemoral joint. Knee Surgery, Sports Traumatology, Arthroscopy. 2014;**22**:2522-2530

[29] Gigante A, Enea D, Greco F, et al. Distal realignment and patellar autologous chondrocyte implantation: Mid-term results in a selected population. Knee Surgery, Sports Traumatology, Arthroscopy. 2009;**17**:2-10

[30] Bauer S, Khan RJ, Ebert JR, et al. Knee joint preservation with combined neutralising high tibial osteotomy (HTO) and Matrix-induced Autologous Chondrocyte Implantation (MACI) in younger patients with medial knee osteoarthritis: A case series with prospective clinical and MRI follow-up over 5 years. The Knee. 2012;**19**:431-439

[31] Vijayan S, Bartlett W, Bentley G, et al. Autologous chondrocyte implantation for osteochondral lesions in the knee using a bilayer collagen membrane and bone graft: A two- to eight-year follow-up study. Journal of Bone and Joint Surgery. British Volume (London). 2012;**94**:488-492

[32] Crawford DC, Heveran CM, Cannon WD, Foo LF, Potter HG. An autologous cartilage tissue implant NeoCart for treatment of grade III chondral injury to the distal femur: Prospective clinical safety trial at 2 years. The American Journal of Sports Medicine. 2009;**37**:1334-1343

[33] Crawford DC, DeBerardino TM, Williams RJ. NeoCart, an autologous cartilage tissue implant, compared with microfracture for treatment of distal femoral cartilage lesions: An FDA phase-II prospective, randomized clinical trial after two years. The Journal of Bone and Joint Surgery. American Volume. 2012;**94**:979-989

[34] Anderson DE, Williams RJ, DeBerardino TM, et al. Magnetic resonance imaging characterization and clinical outcomes after NeoCart surgical therapy as a primary reparative treatment for knee cartilage injuries. The American Journal of Sports Medicine. 2017;**45**:875-883

[35] Flohé S, Betsch M, Ruße K, Wild M, Windolf J, Schulz M. Comparison of two different matrix-based autologous chondrocyte transplantation systems: 1 year follow-up results. European Journal of Trauma and Emergency Surgery. 2011;**37**:397-403

[36] Petri M, Broese M, Simon A, et al. CaReS (MACT) versus microfracture in treating symptomatic patellofemoral cartilage defects: A retrospective matched-pair analysis. Journal of Orthopaedic Science. 2013;**18**:38-44

[37] Schneider U, Rackwitz L, Andereya S, et al. A prospective multicenter study on the outcome of type I collagen hydrogel-based autologous chondrocyte implantation (CaReS) for the repair of articular cartilage defects in the knee. The American Journal of Sports Medicine. 2011;**39**:2558-2565

[38] Welsch GH, Mamisch TC, Zak L, et al. Evaluation of cartilage repair tissue after matrix-associated autologous chondrocyte transplantation using a hyaluronic-based or a collagen-based scaffold with morphological MOCART scoring and biochemical T2 mapping: Preliminary results. The American Journal of Sports Medicine. 2010;**38**:934-942

[39] Zak L, Albrecht C, Wondrasch B, et al. Results 2 years after matrix-associated autologous chondrocyte transplantation using the Novocart 3D scaffold: An analysis of clinical and radiological data. The American Journal of Sports Medicine. 2014;**42**:1618-1627

[40] Niethammer TR, Safi E, Ficklscherer A, et al. Graft maturation of autologous chondrocyte implantation: Magnetic resonance investigation with T2 mapping. The American Journal of Sports Medicine. 2014;**42**:2199-2204

[41] Niethammer TR, Pietschmann MF, Ficklscherer A, Gülecyüz MF, Hammerschmid F, Müller PE. Incomplete defect filling after third generation autologous chondrocyte implantation. Archives of Medical Science. 2016;**12**:785-792

[42] Niethammer TR, Pietschmann MF, Horng A, et al. Graft hypertrophy of matrix-based autologous chondrocyte implantation: A two-year follow-up study of NOVOCART 3D implantation in the knee. Knee Surgery, Sports Traumatology, Arthroscopy. 2014;**22**:1329-1336

[43] Panagopoulos A, van Niekerk L, Triantafillopoulos I. Autologous chondrocyte implantation for knee cartilage injuries: Moderate functional outcome and performance in patients with high-impact activities. Orthopedics. 2012;**35**:e6-14

[44] Niethammer TR, Holzgruber M, Gülecyüz MF, Weber P, Pietschmann MF, Müller PE. Matrix based autologous chondrocyte implantation in children and adolescents: A match paired analysis in a follow-up over three years post-operation. International Orthopaedics. 2017;**41**:343-350

[45] Angele P, Fritz J, Albrecht D, Koh J, Zellner J. Defect type, localization and marker gene expression determines early adverse events of matrix-associated autologous chondrocyte implantation. Injury. 2015;**46**(Suppl 4):S2-S9

[46] Niethammer TR, Müller PE, Safi E, et al. Early resumption of physical activities leads to inferior clinical outcomes after matrix-based autologous chondrocyte implantation in the knee. Knee Surgery, Sports Traumatology, Arthroscopy. 2014;**22**:1345-1352

[47] Kon E, Filardo G, Berruto M, et al. Articular cartilage treatment in high-level male soccer players: A prospective comparative study of arthroscopic second-generation autologous chondrocyte implantation versus microfracture. The American Journal of Sports Medicine. 2011;**39**:2549-2557

[48] Kon E, Filardo G, Condello V, et al. Second-generation autologous chondrocyte implantation: Results in patients older than 40 years. The American Journal of Sports Medicine. 2011;**39**:1668-1675

[49] Ferruzzi A, Buda R, Faldini C, et al. Autologous chondrocyte implantation in the knee joint: Open compared with arthroscopic technique. Comparison at a minimum follow-up of five years. The Journal of Bone and Joint Surgery. American Volume. 2008;**90**(Suppl 4): 90-101

[50] Nehrer S, Chiari C, Domayer S, Barkay H, Yayon A. Results of chondrocyte implantation with a fibrin-hyaluronan matrix: A preliminary study. Clinical Orthopaedics and Related Research. 2008;**466**:1849-1855

[51] Eshed I, Trattnig S, Sharon M, et al. Assessment of cartilage repair after chondrocyte transplantation with a fibrin-hyaluronan matrix—Correlation of morphological MRI, biochemical T2 mapping and clinical outcome. European Journal of Radiology. 2012;**81**:1216-1223

[52] Choi NY, Kim BW, Yeo WJ, et al. Gel-type autologous chondrocyte (Chondron) implantation for treatment of articular cartilage defects of the knee. BMC Musculoskeletal Disorders. 2010;**11**:103

[53] Kim MK, Choi SW, Kim SR, Oh IS, Won MH. Autologous chondrocyte implantation in the knee using fibrin. Knee Surgery, Sports Traumatology, Arthroscopy. 2010;**18**:528-534

[54] Könst YE, Benink RJ, Veldstra R, van der Krieke TJ, Helder MN, van Royen BJ. Treatment of severe osteochondral defects of the knee by combined autologous bone grafting and autologous chondrocyte implantation using fibrin gel. Knee Surgery, Sports Traumatology, Arthroscopy. 2012;**20**:2263-2269

[55] Clavé A, Potel JF, Servien E, Neyret P, Dubrana F, Stindel E. Third-generation autologous chondrocyte implantation versus mosaicplasty for knee cartilage injury: 2-year randomized trial. Journal of Orthopaedic Research. 2016;**34**:658-665

[56] Selmi TA, Verdonk P, Chambat P, et al. Autologous chondrocyte implantation in a novel alginate-agarose hydrogel: Outcome at two years. Journal of Bone and Joint Surgery. British Volume (London). 2008;**90**:597-604

[57] Tohyama H, Yasuda K, Minami A, et al. Atelocollagen-associated autologous chondrocyte implantation for the repair of chondral defects of the knee: A prospective multicenter clinical trial in Japan. Journal of Orthopaedic Science. 2009;**14**:579-588

[58] Tadenuma T, Uchio Y, Kumahashi N, et al. Delayed gadolinium-enhanced MRI of cartilage and T2 mapping for evaluation of reparative cartilage-like tissue after autologous chondrocyte implantation associated with Atelocollagen-based scaffold in the knee. Skeletal Radiology. 2016;**45**:1357-1363

[59] Ossendorf C, Kaps C, Kreuz PC, Burmester GR, Sittinger M, Erggelet C. Treatment of posttraumatic and focal osteoarthritic cartilage defects of the knee with autologous polymer-based three-dimensional chondrocyte grafts: 2-year clinical results. Arthritis Research & Therapy. 2007;**9**:R41

[60] Erggelet C, Kreuz PC, Mrosek EH, et al. Autologous chondrocyte implantation versus ACI using 3D-bioresorbable graft for the treatment of large full-thickness cartilage lesions of the knee. Archives of Orthopaedic and Trauma Surgery. 2010;**130**:957-964

[61] Zeifang F, Oberle D, Nierhoff C, Richter W, Moradi B, Schmitt H. Autologous chondrocyte implantation using the original periosteum-cover technique versus matrix-associated autologous chondrocyte implantation: A randomized clinical trial. The American Journal of Sports Medicine. 2010;**38**:924-933

[62] Kreuz PC, Müller S, Freymann U, et al. Repair of focal cartilage defects with scaffold-assisted autologous chondrocyte grafts: Clinical and biomechanical results 48 months after transplantation. The American Journal of Sports Medicine. 2011;**39**:1697-1705

[63] Kreuz PC, Müller S, Ossendorf C, Kaps C, Erggelet C. Treatment of focal degenerative cartilage defects with polymer-based autologous chondrocyte grafts: Four-year clinical results. Arthritis Research & Therapy. 2009;**11**:R33

Chapter 9

Management of Knee Cartilage Defects with the Autologous Matrix-Induced Chondrogenesis (AMIC) Technique

Michael E. Hantes and Apostolos H. Fyllos

Additional information is available at the end of the chapter

http://dx.doi.org/10.5772/intechopen.71776

Abstract

The arthroscopic findings of knee articular cartilage lesions are reported to be as high as 60%, although only a fragment of these are considered to be symptomatic. Such lesions are believed to accelerate the onset of arthritis. Long-term results of the microfracture technique for chondral and osteochondral defects of the knee cartilage are not satisfactory. The autologous matrix induced chondrogenesis (AMIC) technique offers a promising alternative as an effective cartilage repair procedure in the knee resulting in stable clinical results and with a wide range of indications. An extensive literature review has been performed aiming at providing the rationale behind AMIC, to report clinical results of AMIC and to compare AMIC with other chondrogenesis techniques. Finally, we comment on the appropriate surgical technique and its indications, since the number of one-step arthroscopic procedure proposals is steadily increasing.

Keywords: matrix-induced chondrogenesis, cartilage, microfractures, AMIC

1. Introduction

Despite its durable mechanical properties, hyaline cartilage has low intrinsic regenerative and reparative capacity since it lacks blood supply, nerves and lymphangion. Cartilage defects potentially lead to severe osteoarthritis and disability, and painful symptomatology during that process. None of the pharmacological or surgical cartilage degeneration management options have clearly shown the potential of restoring chondral surface, in order to avoid prosthetic replacement in the final stages of the disease. Numerous reparative techniques for resurfacing articular cartilage defects are currently under extensive clinical research, with promising results. These include cell-based and cell-free materials such as autologous and allogeneic cell-based approaches and multipotent stem-cell-based techniques [1].

Microfracture technique, a low-cost, fully arthroscopic procedure, is still considered the gold standard approach for small cartilage lesions (less than 2 cm^2), not without dispute. This technique enhances migration of mesenchymal stem cells (MSCs) from bone marrow bleeding to the site of a cartilage defect; however, it often results in the formation of fibrocartilage that is biochemically and biomechanically inferior to hyaline articular cartilage, not to mention scatter of the newly formed blood clot into the joint [2]. Its efficacy as a marrow stimulating technique is being questioned due to progressive decrease of the clinical benefit after 2 years, especially as far as large defects are concerned [3].

The autologous matrix induced chondrogenesis (AMIC) offers a promising alternative as an effective cartilage repair procedure in the knee resulting in stable clinical results. It is a one-step procedure that combines microfracture with the application of a biological scaffold acting as a collagen, cell-free matrix that covers the produced blood clot, permitting the containment and ingrowing of MSCs to differentiate into the chondrogenic lineage. The clot induces a repair that covers the cartilage defect with a combination of fibrous and hyaline-like cartilage. AMIC has the potential for homogeneous distribution of MSCs under the membrane that could enhance chondrogenesis and accelerate cartilage healing.

2. Incidence, symptomatology, diagnosis and classification of chondral lesions

Chondral lesions are caused through degradation of joint cartilage, in response to metabolic, genetic, vascular or traumatic stimuli. Chondral defects have been macroscopically graded by the International Cartilage Repair Society (ICRS) in a systematic manner, a system with good inter- and intra-observer reliability [4]. Most commonly used classification systems are the ICRS system and the modified Outerbridge.

The real incidence of osteochondral lesions in humans is unknown, because a large proportion of them are asymptomatic or undiagnosed. The prevalence of single or multiple focal knee articular cartilage pathologies (excluding osteoarthritis and chondromalacia patellae) is reported as high as 30% in arthroscopies, the commonest sites being the medial femoral condyle and the patella [5, 6]. 60% of the lesions are considered to be as severe as grade 3 or worse according to the ICRS system, while 64% of all chondral lesions have a diameter of less than 1 cm [6, 7]. Medial meniscus tears (37%) and ACL ruptures (36%) are the most common concomitant injuries with articular cartilage injuries. The presence of other injuries further influences management of these lesions, such as ACL insufficiency, patellar instability and patellofemoral malalignment [8, 9].

Patients with articular cartilage injuries usually complain of arthritis-like symptoms, such as pain, effusion, and mechanical symptoms varying with location of the lesion. Patients' history is important, although only 60% of patients with a chondral defect diagnosis definitely correlate their symptoms with a specific traumatic incidence [6]. Clinicians should consider the patients' age and presence of a meniscal tear for the odds of having a chondral lesion subsequent to having an ACL injury. Advanced patient's age and long time from initial ACL injury are predictive factors of the severity of chondral lesions, and time from initial ACL injury is significantly associated with the number of chondral lesions [8–10]. However, no reliable correlation between clinical symptoms and articular cartilage status has been established.

Appropriate imaging modality to reach diagnosis is cartilage-sensitive MRI, but definitive diagnosis and classification is set by arthroscopy. Cartilage is a soft, viscoelastic tissue with strong imaging and anisotropic mechanical properties. The MRI signal properties are dependent on the cellular composition of collagen, proteoglycan, and water, but also the MR pulse sequence utilized. Normal cartilage demonstrates "gray-scale stratification," with lower signal intensity closer to the tidemark and subchondral plate and higher signal intensity in the transitional zone, related largely to collagen orientation in the extracellular matrix. Loss of normal gray-scale stratification is an important clinical feature that may herald subsequent delamination of cartilage from the subchondral bone. The assessment and grading of chondral and osteochondral injuries by using MR imaging are straightforward when true morphologic alterations are present. In the setting of higher grade acute injury, the signal alteration in the articular cartilage is readily visible and frequently associated with altered signal intensity in the adjacent subchondral bone marrow and displaced cartilage. On the other hand, low-grade chondral injury typically involves very little morphologic change. Traditional grading systems have classically used altered T2 signal within the cartilage to infer the presence of infrastructural damage. Recent developments in quantitative MR imaging provides direct evaluation of tissue biochemistry in the setting of injury. Several techniques are available to assess the integrity of cartilage glycosaminoglycan, including sodium MR imaging, delayed gadolinium-enhanced MR imaging of cartilage, and T1 ϱ imaging. To assess collagen orientation, quantitative T2 mapping is most often utilized [11, 12].

3. Indications

Operative treatment for chondral and osteochondral knee defects generally is unavoidable at some point and is indicated when nonoperative symptomatic methods fail to relieve pain and mechanical symptoms. Treatment options include debridement, marrow stimulation, transplantation to fill the defect, cell-based therapy, and the use of growth factors or pharmacological agents. The choice of procedure is based primarily on the classification of the lesion and the activity demands of the patient. AMIC is a fairly new but very promising method for cartilage regeneration and was first described by Behrens and Benthien in 2011 [13]. It is a one-step and culture-free procedure, it has the potential for homogeneous distribution of chondrocytes and MSCs to enhance chondrogenesis, and it also has the ability to regenerate hyaline-like cartilage tissue. It has been proven that MSCs can be isolated from the matrix material [14]. The literature currently supports AMIC procedures for moderate to large (greater than 6 cm^2) full thickness defects in the high-demand (but also highly compliant) young patient [15]. Some authors have mentioned underlying rheumatic disease and total meniscectomy as contraindications, whereas "kissing" lesions are unanimously considered an absolute contraindication. Needless to say that elderly patients (although the age limit is not yet determined) with advanced osteoarthritis and significant narrowing of the joint lines are more suitable for total knee arthroplasty than AMIC or similar to AMIC joint preserving interventions.

4. The basic science behind AMIC

In vivo signaling molecules and biomechanical stimuli provides a much more appropriate environment for progenitor cells to differentiate than in vitro chondrogenesis. Fibrin, PDGF and other factors contained in a natural blood clot are highly chemoattractive for MSC. PDGF-BB, EGF and TGF-b are the most important potent mitogens. These factors also contained in the blood clot after subchondral microfracture induce the migration of MSC and have at the same time the potency to enhance their proliferation. Therefore, the migration and proliferation steps of MSC can take place simultaneously in vivo, excluding the need for in vitro culturing. Furthermore, cartilage differentiation initiates in contact with subchondral bone and earliest chondrogenesis is often seen in areas where active remodeling of the subchondral bone plate occurs and, thus, enhanced nutrition and a higher anabolic rate of the cells can take place. MSCs derived from microfractures have the same phenotypic plasticity as chondrogenic cells in the cartilage basal zone. One cm^3 of blood from a single microfracture hole has approximately 8000 CD34+ MSCs [16, 17].

Strength of integration of the neotissue depends on the age and metabolic activity of the tissue. The use of more immature cells has obvious benefits for integration, which argues in favor of MSC-based as opposed to chondrocyte-based repair strategies. Collagen and fibrin-based gels are subject to strong shrinking during chondrogenesis which points towards an increasing risk of partial defect filling and loss of a superclot after microfracturing during progress of chondrogenic differentiation. To be able to avoid this, a clinically applied solid collagen type I/III matrix as used in the AMIC technique appears to facilitate chondrogenesis of MSC. It has been proven that bone marrow cells can be guided directly to a cartilage defect by a collagenous matrix and MSCs can be isolated regularly from the matrix [14]. Inhibitory signals may come from the opposed cartilage surface and synovial fluid to dominate the surface area of fibrous repair tissue. The lowest cartilage layer is responsible for load transmission from cartilage into bone. Application of biomechanical loading during chondrogenesis of MSC stimulated cartilaginous matrix production in tissue engineering applications underlining the importance of mechanical signals for tissue guidance during repair [17].

5. Surgical technique

The AMIC procedure can be performed with either a mini open approach, or a combination of arthroscopy and mini arthrotomy, or even as an all-arthroscopic technique. There are different types of scaffolds available: natural protein–based or carbohydrate-based scaffolds, and synthetic scaffolds. The 3 scaffolds that have been reported in the literature for AMIC are ChondroGide (Geistlich Biomaterials, Wolhausen, Switzerland), Hyalofast (Fidia Advanced Biopolymers, Padua, Italy), and Chondrotissue (BioTissue, Zurich, Switzerland) [18]. Modifications and enrichment of the scaffold with newer biomaterials (such as platelet-rich plasma or leucocyte-platelet-concentrated membrane) of the original AMIC technique may improve cartilage repair outcome and optimize the operative approach [19]. The basic procedural rationale is chondral defect arthroscopic debridement and preparation of smooth surrounding boundaries, followed by subchondral microfracture technique and finally stable bilayer matrix fixation.

The patient is placed supine in an ordinary arthroscopy setup, under regional or general anesthesia, antibiotic prophylaxis and with tourniquet application to the thigh. The status of the joint, ligamentous structure integrity and the cartilage lesion are assessed by arthroscopy, including location, size, and depth according to the ICRS classification. Clear, smooth and stable borders of normal adjacent cartilage are defined, followed by removal of the calcified chondral layer with a burr or a curette. According to the original technique, a mini arthrotomy is performed at this stage and access to the subchondral bone is reached by nanofractures or microfractures or by microdrilling with appropriate instruments. The gaps between the holes should permit sufficient bridging to prevent subchondral fracture. Generally, nanofractures technique is preferred, with standardized drilled holes nine millimeters deep and one millimeter in diameter and standard needle angling [20–22]. All-arthroscopic techniques have been described as well [22–24]. The collagen matrix of choice is consequently trimmed to fill the size of the defect, usually by template or imprint. Undersize of the scaffold is recommended to avoid dislocation with movement. The matrix is then pressed and sutured or glued (allogenic or partially autologous fibrin glue) or with a combination of both stabilized on the defect, making sure that a smooth transition to normal cartilage has been ensured. There are usually two sides of the membrane; the rough side faces the subchondral bone and the smooth side faces the joint. The application of fibrin glue and the attachment of the membrane is best done in a dry environment in case of all-arthroscopic technique. The scaffold acts like a sponge that holds the blood clot within the defect and induces hemostasis while protecting the underlying tissue. This may be either performed arthroscopically or as an arthroscopically assisted mini-open technique. Before closure, multiple gentle movements of the joint are advised in order to confirm unhindered membrane placement. Irrigation of the joint is discouraged as this may almost certainly result in membrane dislocation and removal of the desired blood clot. The use of a drain deems unnecessary, not to mention that suction could result in membrane dislodgement (**Figures 1–9**).

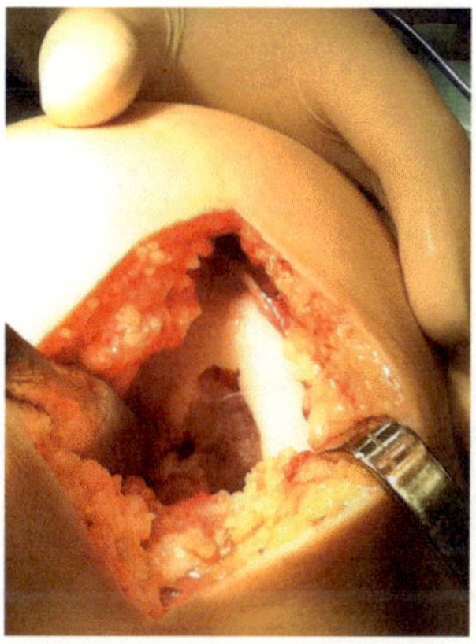

Figure 1. Mini arthotomy, identification and classification of the lesion.

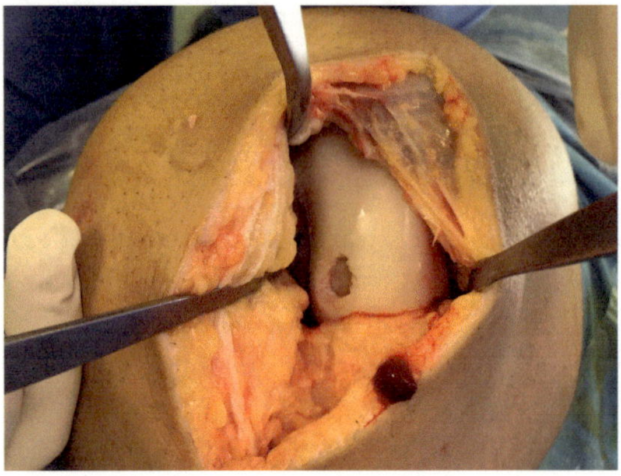

Figure 2. Osteochondral lesion (ICRS grade 4) after open debridement and preparation of boundaries.

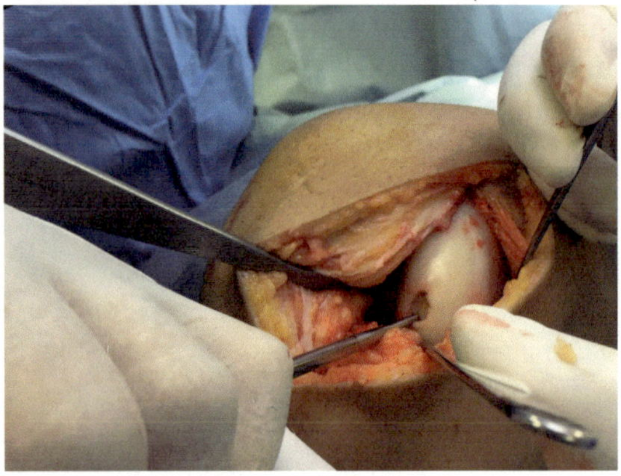

Figure 3. Performing nanofractures.

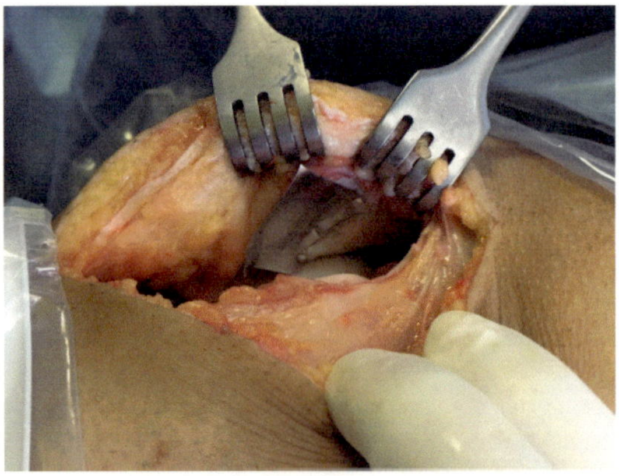

Figure 4. Imprint technique with aluminum foil.

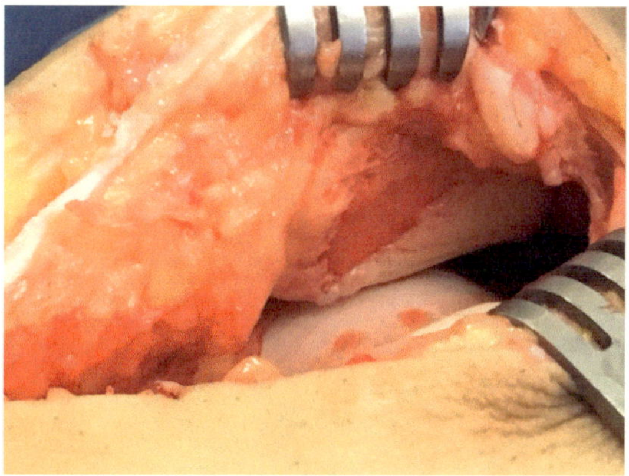

Figure 5. After open scaffold placement in a large patellar defect.

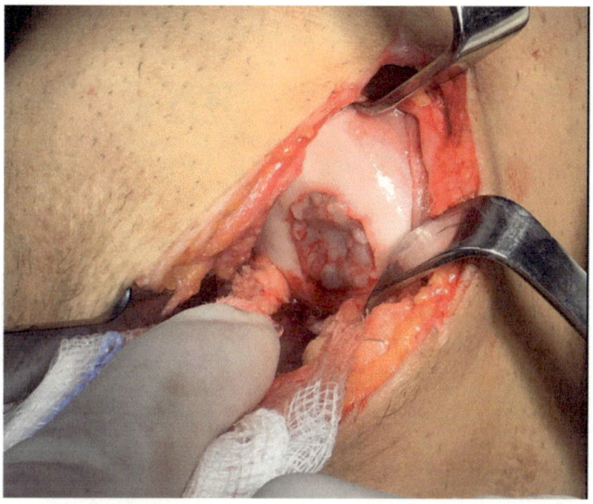

Figure 6. After open membrane placement in medial femoral condyle osteochondral lesion.

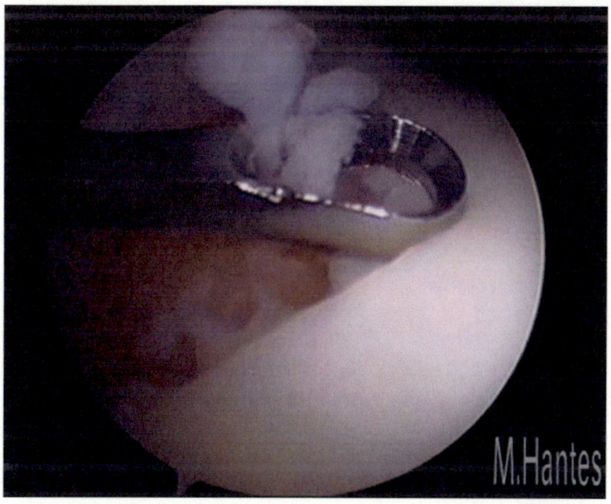

Figure 7. Arthroscopic curettage of osteochondral lesion to healthy bone depth.

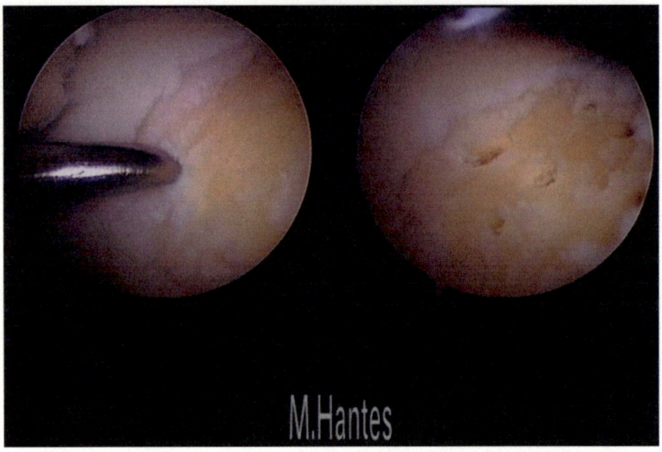

Figure 8. Arthroscopic microfracture technique.

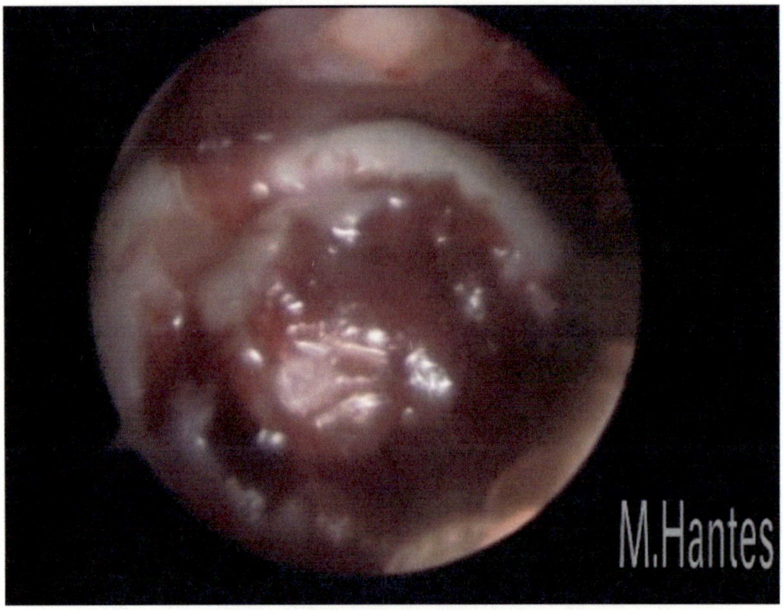

Figure 9. Membrane attached after preparation of osteochondral defect under dry arthroscopy.

6. Rehabilitation

Patient compliance is the key for success of this sensitive procedure, although consensus has not been reached. Most authors recommend foot sole contact for 6 weeks using crutches building up full weight bearing after 8 weeks. Partial weight bearing pertains to the possible risk of a compression fracture after microfracturing due to small and ill-defined bone bridges which might not bear enough weight. Articular remodeling and chondral maturation may take up to

6 months so limited weight bearing for a certain amount of time is important. However, remodeling of the chondral matrix may actually profit from early mobilization using a combination of compression and shear forces. Since there is sufficient bridging between the drilled holes and the holes are straight there should be no reason for a subchondral impression fracture. Earlier weight-bearing has been suggested after nanofractures [25–27].

Range of motion is generally restricted for 6 weeks depending on site of the lesion. When the femoral condyles are involved, active and passive knee flexion up to 90° is permitted, whereas when the patella is involved knee flexion is restricted to 60° for the first month. Mobilization exercises including continuous passive motion, electrotherapy of leg muscles and proprioception training are an integral part of the rehabilitation program. Unrestricted weight-bearing and range of motion is permitted after 8 weeks. Contact sports are prohibited for at least a year [28–30]. No study has yet addressed the effect of rehabilitation on the quality of the repair.

7. Results

Well-established rating systems have been used to summarize relevant outcome measures. The combination of the Lysholm score and the Visual Analog Score (VAS) have been recommended in the literature before. The Lysholm scoring system has demonstrated validity, reliability and responsiveness to cartilage pathology and treatment. The VAS has widely been used to monitor subjective satisfaction postoperatively [31, 32]. The Modified Cincinnati, the Modified ICRS scores and the knee injury and osteoarthritis outcome score (KOOS) have also been suggested.

Structural repair can be assessed with MRI with a focus on the extent, signal intensity, and surface of the defect filling, integration to adjacent cartilage, and bone marrow lesion. Semiquantitative MRI scores of osteoarthritis established as BLOKS and WORMS can play the part. But it is magnetic resonance observation of cartilage repair tissue (MOCART) protocol that is more often used, with almost perfect interobserver reliability. The detection of subtle cartilage changes by MRI requires high resolution imaging, which is not provided by standard sequences. With the use of a surface coil placed over the knee compartment of interest, high resolution imaging with reasonable scan times is possible on routinely used 1 or 1.5 T MRI units by performing fast spin-echo imaging. The advantage of this imaging technique is that it can be used on every standard 1 or 1.5 T scanner. Nine variables are used to describe the morphology and signal intensity of the repair tissue compared to the adjacent native cartilage, according to the MOCART system. A statistically significant correlation between the clinical outcome (as measured by VAS and KOOS) and some of the radiological variables, including the filling of the defect, the structure of the repair tissue, changes in the subchondral bone and the signal intensities has been established [33] (**Figure 10**).

Encouraging mid- to long-term results have been published that make the AMIC procedure seem promising for a wide range of indications. Gille et al. published their results after 2 years of follow-up of 57 patients who had undergone AMIC (and concomitant procedures in appropriate cases). Mean defect size was 3.4 cm^2 and classification grade in the Outerbridge system was III or higher, with mean patient age of 37 years. Mean Lysholm and VAS scores were

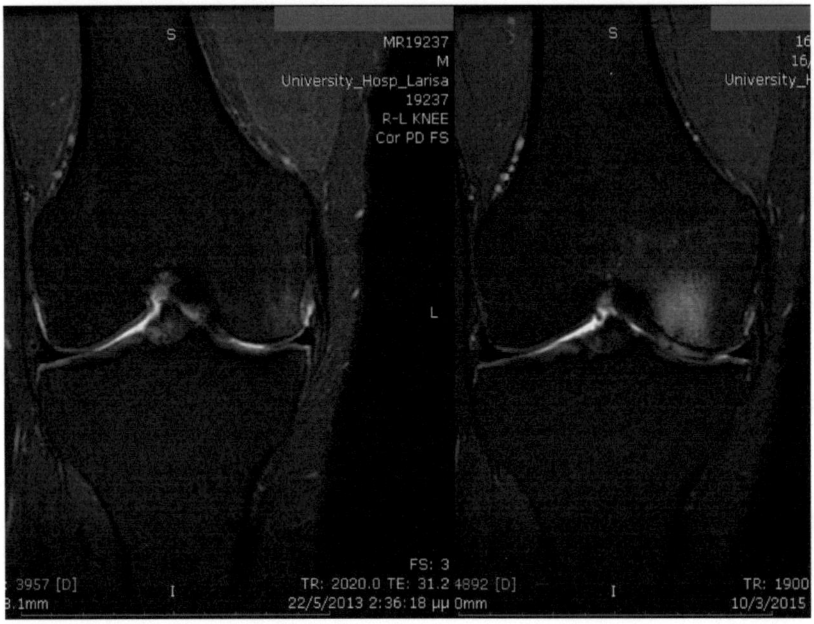

Figure 10. MRI of patient pre- and 19 months post-op, with good filling of the chondral defect in the medial femoral condyle and good integration of the neotissue.

significantly improved in all age groups at 2 years post-op. Defect size (range 0–12 cm^2) had no impact on the clinical outcome and no adverse effects or procedural failures were reported [34]. Furthermore, another prospective randomized control trial of 47 patients with mean age 37 years and mean defect size 3.6 cm^2, directly compared results after microfracturing alone with results after AMIC. After improvement for the first 2 years in all sub-groups, a progressive and significant score degradation was observed in the microfracture group, while all functional parameters remained stable for at least 5 years in the AMIC group. At two and 5 years, MRI defect filling was more complete in the AMIC groups (at least 60% of the patients had a defect filling of more than 2/3). No serious treatment-related adverse events were reported. Biopsies were obtained at 2 years in two patients, both belonging to the AMIC group. Both showed the presence of a fibrocartilaginous matrix, without evidence of residual membrane material, and in one case cell cluster formation was observed in the deep zone of the repair tissue. Hyaline cartilage specific markers were identified, as Safranin-O, collagen-type I and II and a glycosaminoglycan. Both repair tissues were characterized as mostly fibrocartilaginous [28]. In a retrospective review of results in 40 knees with a mean follow-up of 28.8 months, AMIC alone and in combination with unloading osteotomy or patella realignment significantly improved symptomatic knees with isolated osteochondral and chondral lesions. A relatively important complication rose, knee stiffness in the subgroup with patella defects, and manipulation under anesthesia was necessary. However, subgroups varied considerably in lesion site and size, the patient population was small and radiological results according to the MOCART system were inconsistent and therefore unreliable [29]. Finally, in a prospective trial of 21 patients treated for full-thickness defects larger than 2 cm^2, annual clinical reviews and an MRI was performed at 1 and 7 years postoperatively. All patients showed maintenance of good clinical and functional results 7 years after the procedure, although imaging findings did not support good clinical outcomes in all cases [30].

Two recent meta-analyses pointed out that conclusive evidence to determine whether morphological MRI is reliable in predicting clinical outcome after cartilage repair is lacking. These reports also stated that no MRI classification has been shown to correlate with clinical outcomes after different types of cartilage repair, although AMIC was not among the procedures included in the studies [35, 36]. Since the interpretation of cartilage structure from morphological MRI data is still debated and does not correlate with clinical scores, clinical and functional results should be considered as the most important outcomes, and so far, AMIC shows great clinical benefit for the patient. Finally, it should be outlined that no randomized controlled studies have been published comparing AMIC results with other cartilage repair procedures (apart from microfractures), such as autologous chondrocyte implantation (ACI) or matrix-induced chondrocyte implantation (MACI), in order to draw definite conclusions. The obvious advantage is the fact that it is a one-step procedure, faster, simpler and at a lower cost compared to ACI/MACI, since no cell culture and/or reoperation is needed. Standardization of the AMIC technique is also an issue due to different micro- or nanofracturing technique and open vs. arthroscopic procedures.

8. Conclusion

To sum up, AMIC is a one-step cartilage repair technique performed either by arthroscopy or by mini arthrotomy in the stable and well aligned knee. It shows great promise with good functional mid- and long-term results, and has a very low rate and range of complications and failures. The procedure seems to augment the healing potential thanks to homogeneous distribution of MSCs that enhances chondrogenesis, and also shows ability to regenerate hyaline-like cartilage tissue on MRI. Prospective long-term randomized trials are needed to compare the results of the AMIC procedure with other cartilage repair techniques, as well as to ensure maintenance of good clinical outcomes in the long run. A systematic and prolonged rehabilitation program is essential and outcome is absolutely dependent on patient compliance.

Author details

Michael E. Hantes[1]* and Apostolos H. Fyllos[1,2]

*Address all correspondence to: hantesmi@otenet.gr

1 Department of Orthopedics, Faculty of Medicine, University of Thessaly, Larisa, Greece

2 Department of Anatomy, Faculty of Medicine, University of Thessaly, Larisa, Greece

References

[1] Makris EA, Gomoll AH, Malizos KN, Hu JC, Athanasiou KA. Repair and tissue engineering techniques for articular cartilage. Nature Reviews Rheumatology. 2015 Jan;**11**(1):21-34

[2] Kreuz PC, Steinwachs MR, Erggelet C, Krause SJ, Konrad G, Uhl M, et al. Results after microfracture of full-thickness chondral defects in different compartments in the knee. Osteoarthritis and Cartilage. 2006 Nov;**14**(11):1119-1125

[3] Mithoefer K, McAdams T, Williams RJ, Kreuz PC, Mandelbaum BR. Clinical efficacy of the microfracture technique for articular cartilage repair in the knee: An evidence-based systematic analysis. The American Journal of Sports Medicine. 2009 Oct;**37**(10):2053-2063

[4] Dwyer T, Martin CR, Kendra R, Sermer C, Chahal J, Ogilvie-Harris D, et al. Reliability and validity of the arthroscopic international cartilage repair society classification system: Correlation with histological assessment of depth. Arthroscopy. 2017 Jun;**33**(6):1219-1224

[5] Widuchowski W, Widuchowski J, Trzaska T. Articular cartilage defects: Study of 25,124 knee arthroscopies. The Knee. 2007 Jun;**14**(3):177-182

[6] Hjelle K, Solheim E, Strand T, Muri R, Brittberg M. Articular cartilage defects in 1,000 knee arthroscopies. Arthroscopy. 2002 Sep;**18**(7):730-734

[7] Tandogan RN, Taser O, Kayaalp A, Taşkiran E, Pinar H, Alparslan B, et al. Analysis of meniscal and chondral lesions accompanying anterior cruciate ligament tears: Relationship with age, time from injury, and level of sport. Knee Surgery, Sports Traumatology, Arthroscopy. 2004 Jul;**12**(4):262-270

[8] Logerstedt DS, Snyder-Mackler L, Ritter RC, Axe MJ. Knee pain and mobility impairments: Meniscal and articular cartilage lesions. The Journal of Orthopaedic and Sports Physical Therapy. 2010 Jun;**40**(6):A1-A35

[9] Michalitsis S, Vlychou M, Malizos KN, Thriskos P, Hantes ME. Meniscal and articular cartilage lesions in the anterior cruciate ligament-deficient knee: Correlation between time from injury and knee scores. Knee Surgery, Sports Traumatology, Arthroscopy. 2015 Jan;**23**(1):232-239

[10] Flanigan DC, Harris JD, Trinh TQ, Siston RA, Brophy RH. Prevalence of chondral defects in athletes' knees: A systematic review. Medicine and Science in Sports and Exercise. 2010 Oct;**42**(10):1795-1801

[11] Pathria MN, Chung CB, Resnick DL. Acute and stress-related injuries of bone and cartilage: Pertinent anatomy, basic biomechanics, and imaging perspective. Radiology. 2016 Jul;**280**(1):21-38

[12] Potter HG, Koff MF. MR imaging tools to assess cartilage and joint structures. HSS Journal. 2012 Feb;**8**(1):29-32

[13] Benthien JP, Behrens P. The treatment of chondral and osteochondral defects of the knee with autologous matrix-induced chondrogenesis (AMIC): Method description and recent developments. Knee Surgery, Sports Traumatology, Arthroscopy. 2011 Aug;**19**(8):1316-1319

[14] Kramer J, Böhrnsen F, Lindner U, Behrens P, Schlenke P, Rohwedel J. In vivo matrix-guided human mesenchymal stem cells. Cellular and Molecular Life Sciences. 2006 Mar;**63**(5): 616-626

[15] Zhang C, Cai YZ, Lin XJ. One-step cartilage repair technique as a next generation of cell therapy for cartilage defects: Biological characteristics, preclinical application, surgical techniques, and clinical developments. Arthroscopy. 2016 Jul;**32**(7):1444-1450

[16] Tallheden T, Dennis JE, Lennon DP, Sjögren-Jansson E, Caplan AI, Lindahl A. Phenotypic plasticity of human articular chondrocytes. The Journal of Bone and Joint Surgery. American Volume. 2003;**85-A**(Suppl 2):93-100

[17] Richter W. Mesenchymal stem cells and cartilage in situ regeneration. Journal of Internal Medicine. 2009 Oct;**266**(4):390-405

[18] Lee YH, Suzer F, Thermann H. Autologous matrix-induced chondrogenesis in the knee: A review. Cartilage. 2014 Jul;**5**(3):145-153

[19] D'Antimo C, Biggi F, Borean A, Di Fabio S, Pirola I. Combining a novel leucocyte-platelet-concentrated membrane and an injectable collagen scaffold in a single-step AMIC procedure to treat chondral lesions of the knee: A preliminary retrospective study. European Journal of Orthopaedic Surgery and Traumatology. 2016 Jul;**27**(5):673-681

[20] Steadman JR, Rodkey WG, Rodrigo JJ. Microfracture: Surgical technique and rehabilitation to treat chondral defects. Clinical Orthopaedics and Related Research. 2001 Oct;(391 Suppl):S362-S369

[21] Chen H, Sun J, Hoemann C, Lascau-Coman V, Ouyang W, Trankhanh N, et al. Drilling and microfracture lead to different bone structure and necrosis during bone-marrow stimulation for cartilage repair. Journal of Orthopaedic Research. 2009 Nov;**27**(11):1432-1438

[22] Benthien JP, Behrens P. Nanofractured autologous matrix induced chondrogenesis (NAMIC©)—Further development of collagen membrane aided chondrogenesis combined with subchondral needling: A technical note. The Knee. 2015 Oct;**22**(5):411-415

[23] Piontek T, Ciemniewska-Gorzela K, Szulc A, Naczk J, Słomczykowski M. All-arthroscopic AMIC procedure for repair of cartilage defects of the knee. Knee Surgery, Sports Traumatology, Arthroscopy. 2012 May;**20**(5):922-925

[24] Sadlik B, Wiewiorski M. Implantation of a collagen matrix for an AMIC repair during dry arthroscopy. Knee Surgery, Sports Traumatology, Arthroscopy. 2015 Aug;**23**(8):2349-2352

[25] Benthien JP, Behrens P. Reviewing subchondral cartilage surgery: Considerations for standardized and outcome predictable cartilage remodelling. International Orthopaedics. 2013 Nov;**37**(11):2139-2145

[26] Wang N, Grad S, Stoddart MJ, Niemeyer P, Reising K, Schmal H, et al. Particulate cartilage under bioreactor-induced compression and shear. International Orthopaedics. 2014 May;**38**(5):1105-1111

[27] Schaetti O, Grad S, Goldhahn J, Salzmann G, Li Z, Alini M, et al. A combination of shear and dynamic compression leads to mechanically induced chondrogenesis of human mesenchymal stem cells. European Cells & Materials. 2011 Oct 11;**22**:214-225

[28] Volz M, Schaumburger J, Frick H, Grifka J, Anders S. A randomized controlled trial demonstrating sustained benefit of Autologous Matrix-Induced Chondrogenesis over microfracture at five years. International Orthopaedics. 2017 Apr;**41**(4):797-804

[29] Kusano T, Jakob RP, Gautier E, Magnussen RA, Hoogewoud H, Jacobi M. Treatment of isolated chondral and osteochondral defects in the knee by autologous matrix-induced chondrogenesis (AMIC). Knee Surgery, Sports Traumatology, Arthroscopy. 2012 Oct; **20**(10):2109-2115

[30] Schiavone Panni A, Del Regno C, Mazzitelli G, D'Apolito R, Corona K, Vasso M. Good clinical results with autologous matrix-induced chondrogenesis (Amic) technique in large knee chondral defects. Knee Surgery, Sports Traumatology, Arthroscopy. 2017 Mar 21. [Epub ahead of print]

[31] Flandry F, Hunt JP, Terry GC, Hughston JC. Analysis of subjective knee complaints using visual analog scales. The American Journal of Sports Medicine. 1991 Mar-Apr;**19**(2): 112-118

[32] Fuchs S, Friedrich M. Possible influence of knee scores. Der Unfallchirurg. 2000 Jan;**103**(1): 44-50

[33] Marlovits S, Singer P, Zeller P, Mandl I, Haller J, Trattnig S. Magnetic resonance observation of cartilage repair tissue (MOCART) for the evaluation of autologous chondrocyte transplantation: Determination of interobserver variability and correlation to clinical outcome after 2 years. European Journal of Radiology. 2006 Jan;**57**(1):16-23

[34] Gille J, Behrens P, Volpi P, de Girolamo L, Reiss E, Zoch W, et al. Outcome of Autologous Matrix Induced Chondrogenesis (AMIC) in cartilage knee surgery: Data of the AMIC Registry. Archives of Orthopaedic and Trauma Surgery. 2013 Jan;**133**(1):87-93

[35] de Windt TS, Welsch GH, Brittberg M, Vonk LA, Marlovits S, Trattnig S, Saris DB. Is magnetic resonance imaging reliable in predicting clinical outcome after articular cartilage repair of the knee? A systematic review and meta-analysis. The American Journal of Sports Medicine. 2013 Jul;**41**(7):1695-1702

[36] Blackman AJ, Smith MV, Flanigan DC, Matava MJ, Wright RW, Brophy RH. Correlation between magnetic resonance imaging and clinical outcomes after cartilage repair surgery in the knee: A systematic review and meta-analysis. The American Journal of Sports Medicine. 2013 Jun;**41**(6):1426-1434

Chapter 10

MRI Mapping for Cartilage Repair Follow-up

Mars Mokhtar

Additional information is available at the end of the chapter

http://dx.doi.org/10.5772/intechopen.70372

Abstract

Patients, who benefit from cartilage repair surgery, need a non-invasive and high-quality imaging modality to assess the structure and the biochemical property of the repair tissue. Magnetic resonance imaging (MRI), which provides better tissue contrast and high spatial resolution, is currently the best imaging technique available for the assessment of articular cartilage pathologies. In addition to MR morphology sequences, that allow cartilage lesions detection as well as repair tissue evaluation from the articular surface of the joint to the bone-cartilage interface, MRI mapping techniques help to assess the technical success of the procedure of cartilage repair and the state of cartilage healing, as well the identification of possible complications after cartilage repair surgery. MRI mapping techniques such as T1, T2 and T2* mapping help to assess the biochemical property of the repair tissue using delayed gadolinium-enhanced MRI of cartilage (dGEMRIC) to assess the proteoglycan content and T2/T1rho (T1ϱ) mapping to assess the collagen content and the fiber matrix arrangement. This chapter gives an overview about the MRI mapping techniques used for Cartilage Repair Tissue Follow-up.

Keywords: MRI, cartilage repair, T2 mapping, dGEMRIC, follow-up, T1rho, T2* mapping

1. Introduction

Many techniques are used to evaluate the knee articular cartilage however non-invasive conventional magnetic resonance imaging (MRI) is the method of choice for the evaluation of knee articular cartilage [1]. Imaging of articular cartilage needs MRI sequence which is able to characterize morphological alterations of cartilage as well as adjacent tissue and to measure with high accuracy the cartilage thickness [2]. Conventional MRI sequences allow the detection of degenerative cartilage lesions and the changes due to therapy response, e.g., after cartilage repair procedures.

In addition to the evaluation of cartilage morphology which is possible by MRI conventional 2D or 3D sequences, there is a need to visualize the biochemical components of the cartilage especially after cartilage repair surgery. MRI has been demonstrated to be sensitive to the variation of local water content [3], the loss of collagen content [4] and the organization of the collagen fiber [5] in the extracellular matrix. MRI parameters such as T1, T2 and T2* can serve as marker of biochemical properties of the knee articular cartilage. The most used mapping techniques are T2 and delayed gadolinium-enhanced MRI of cartilage (dGEMRIC). *T*2 mapping was reported to provide information about collagen matrix concentration and organization, whereas dGEMRIC is sensitive to proteoglycan content [6].

2. Cartilage repair surgery techniques

It is very important to know the different repair procedures and the behavior of the repair tissue in MR imaging at various postoperative intervals to evaluate the success of the surgery or to check for any complications [7]. Different methods have been used to stimulate the formation of a new articular cartilage such as microfracture, autologous chondrocyte implantation (ACI) and Osteochondral Allograft.

2.1. Microfracture

This procedure, introduced by Steadman et al., consists of removing all unstable and damaged articular cartilage till the subchondral bone plate, then making multiple small holes in it. This leads to bleeding, clot formation, as well as the introduction of marrow derived stem cells to the site [8]. The microfracture technique is generally used to repair small- to mid-sized cartilage defects in osteoarthritis (OA). It was reported that cartilage tends to deteriorate within a few years [9–11].

2.2. Autologous chondrocyte implantation (ACI)

This technique was first performed by Peterson et al. in Gothenburg in 1987. First, cartilage is harvested from a patient using arthroscopy. Second, it is grown in tissue culture medium. Then, it is reimplanted within the patient's cartilage defect beneath a periosteal patch to produce new cartilage repair tissue [12].

2.3. Osteochondral allograft

Osteochondral allografting involves the replacement of damaged articular cartilage with mature hyaline one from a suitable donor.

3. Cartilage repair surgery follow-up

The ideal cartilage repair tissue should, over time, develop a collagen network with a similar organization and concentration of normal hyaline cartilage [6]. Cartilage repair surgery techniques require a non-invasive postoperative technique to monitor the cartilage repair tissue over time to detect complications or deviation of the normal maturation process. The normal appearance of cartilage repair tissues varies according to the applied surgical technique and the timing of postoperative follow-up.

3.1. Cartilage repair surgery follow-up parameters

Many parameters should be assessed in MR imaging examinations after cartilage repair procedure. Magnetic resonance observation of cartilage repair (MOCART) proposes the assessment of the following MR imaging parameters: the degree of defect repair and defect filling, integration to border zone, quality of repair tissue surface, structure of repair tissue, signal intensity of repair tissue, status of subchondral lamina, integrity of subchondral bone the presence of complications (adhesions and effusion) [13, 14].

This scoring system was validated in a 2-year longitudinal study of patients with matrix assisted chondrocyte implantation and correlated well with clinical scores. The evaluated parameters are the degree of defect filling, structure of repair tissue, change in subchondral bone, and signal intensity of repair tissue [15]. In another study of patients who underwent either microfracture or ACI, the following MR imaging parameters were evaluated: signal intensity relative to native cartilage; morphology with respect to native cartilage; delamination; nature of the interface with the adjacent surface; degree of defect filling; integrity of cartilage on the opposite articular surface and bony hypertrophy [16].

After microfracture and osteochondral autograft transplant sites, MRI can evaluate the degree of defect filling, the extent of integration of repair tissue with adjacent tissues, the presence or absence of proud subchondral bone formation, the characteristics of the graft substance and surface, and the appearance of the underlying bone [7]. After ACI, visualization of the biochemical properties of cartilage becomes more important, since repair tissue shows a gradual maturation over time [17, 18].

3.2. Cartilage repair surgery follow-up timing

In case of articular cartilage repair, first we need to fill the defect area with a tissue that has the same mechanical properties as normal articular cartilage; second we need to promote successful integration between the repair tissue and the native articular cartilage [19]. The parameters which determine the mechanical properties of knee articular cartilage are the content, the arrangement and the interaction between the main components such as the collagen matrix, proteoglycans (PGs), and interstitial water [20]. PGs have been shown to be the primary parameter which determines the compressive properties of cartilage and collagen was reported to responsible for the tensile property [21].

Follow-up MR imaging studies should be performed at 3–6 postoperative months to assess the volume and the integration of repair tissue and after 1 year to evaluate the maturation of the graft and identification of any complications [22]. The ability to evaluate the organization of the collagen matrix in repair tissue over time is important, as failure within the collagenous fiber network is considered as failure of cartilage repair procedure [6].

4. Magnetic resonance imaging (MRI)

The MRI principle can be explained by the fact that atomic nuclei of fluids in a magnetic field can be flipped off their preferred orientation parallel to a magnetic field when exposed to an electromagnetic radio frequency field (RF field). When the RF field is switched off, the atomic nuclei return to their original state and release the absorbed energy as electromagnetic radiation. During excitation, we send radio frequency energy to the hydrogen protons inside the body. Those protons will absorb this energy as a heat. When we stop excitation, the relaxation process starts and the energy introduced during excitation is transferred to the surrounding protons.

There are two types of relaxation. First, the T1 (longitudinal Relaxation) whereby there is energy transfer from the spins to the environment and the T2 (transverse Relaxation) where there is dephasing of spins. The contrast in MRI depends on many parameters mainly patient parameters, sequence type and sequences parameters. The patient related parameters are T1, T2 and proton density. By varying parameters such as repetition time (TR) and echo time (TE), we can obtain weighted sequence like T1, T2 and proton density weighted sequences.

4.1. T1 relaxation

The T1 relaxation curve which describes the relaxation speeds for any given tissue follows an exponential law. The constant T1 is defined as the time required for the longitudinal component of M0 to return to 63% of its initial value. The difference in relaxation times gives the T1 contrast. The T1 value depends on the mass and the size of the molecules constituting the tissue. It depends strongly on B0 and is a function of the micro-viscosity of the medium. For liquid, the values of T1 are greater than the second and for the most structured tissues, the T1 values are of the order of a few 100 ms.

4.2. T2 relaxation

During the T2 relaxation process, each tissue loses transverse coherence (magnetization) via an exponential decay process. T2 is defined as the time after which the transverse magnetization is decayed to 37% of its starting amplitude. T2 is a tissue specific parameter and is weakly dependent on the magnetic field B0 because it happens on a perpendicular plane to B0. In solids, which possess a rigid atomic network, T2 is extremely short, whereas in liquids where the decay of the transverse magnetization takes place slowly, T2 is longer and that is why pure water will appear as hyper signal on a T2-weighted sequence.

5. T1 mapping

The contrasts in MRI morphology sequences depend on the difference of signal intensities between tissues at the time of echo measurement. To display the T1, T2 and T2* values of each tissue, we need to calculate parametric maps. In those maps, the pixel intensities in the image provide quantitative values of the studied relaxation time.

5.1. T1 mapping calculation

To calculate T1 mapping we can use either spin echo or gradient echo sequence. With the 2D spin echo sequence, there are two methods to calculate T1 maps either based on the phase inversion or saturation of the longitudinal magnetization. In each case, at least two data sets with different parameters are needed. In case of spin echo, we need to acquire the same sequences twice with the same parameters but different repetition time (TR) and in case of inversion recovery sequence, we use the same sequence but with different inversion time (TI). The acquisition time required for the T1 mapping using spin echo technique is relatively long and often limited to a small number of slices. 3D gradient echo sequences are better alternative solution which provide high signal to noise ratio (SNR) and thin slices in relatively less acquisition time. 3D spoiled fast gradient echo (3D FLASH) sequence with two different excitation flip angles of was used to assess the T1 relaxation times [23, 24].

5.2. T1 mapping clinical applications

The measurements of T1ρ can be used to visualize interactions between the water molecules in restricted movement and local macromolecular environment. The extracellular matrix of the articular cartilage provides a limited movement environment of water molecules. The modifications of the extracellular matrix, such as loss PG, may be reflected by the change of the T1ϱ values. In one study, the normalized T1ϱ rate was strongly correlated with alterations in fixed charge density (FCD) due to depletion of PG which was confirmed by histology [23].

6. Delayed gadolinium-enhanced MRI of cartilage (dGEMRIC)

A dGEMRIC involves intravenous administration of negatively charged contrast agent ($Gd\text{-}DTPA^{2-}$). After injection of Gd-DTPA2, the contrast agent penetrates the cartilage through both the articular surface and the subchondral bone [24]. Since the contrast agent has negative charge, it will interact with FCD which is directly related to the GAG concentration. The distribution of $Gd\text{-}DTPA^{2-}$ is inversely proportional to glycosaminoglycan (GAG) content of the tissue of interest. T1 relaxation times are inversely proportional to the concentration of $Gd\text{-}DTPA^{2-}$. The $Gd\text{-}DTPA^{2-}$ will shorten the T1 of tissues in this case the cartilage, therefore T1 can be used as a specific marker of GAG concentration. Healthy cartilage, which contains an abundance of GAG, will show a low $Gd\text{-}DTPA^{2-}$ concentration, whereas GAG-depleted degraded cartilage will show a high $Gd\text{-}DTPA^{2-}$ concentration which will result in lower T1 values compared with healthy cartilage [25] (**Figure 1**).

6.1. Exam preparation

It was recommended to inject a bolus of $Gd\text{-}DTPA^{2-}$ with a quantity of 0.2 mmol contrast agent per kilogram body weight (double dose). After injection, we ask the patient to do some exercises of the knee, for example, walking up and down stairs for about 20 minutes. Ninety minutes after IV injection, we acquire the postcontrast MRI study. This delay time of 90 minutes allows the contrast agent to fully diffuse into the cartilage. However, since cartilage thickness is variable within the knee and between patients, the delay time to reach equilibrium has to be adjusted [26]. Moreover, after different cartilage repair surgeries, the timing to reach the equilibrium, and the exercise period are difficult to be defined and standardized [21].

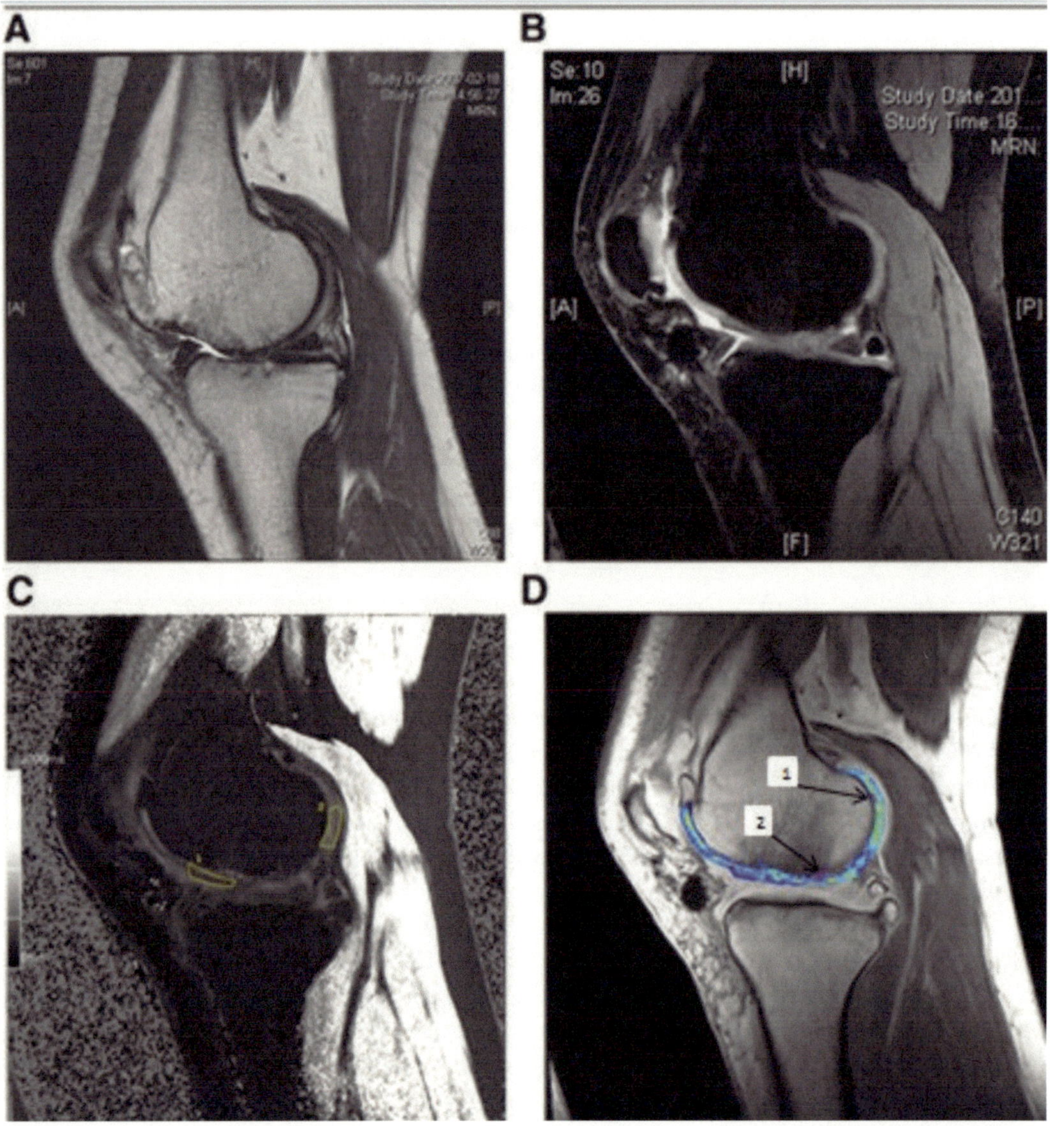

Figure 1. MRI evaluation of cartilage regeneration 3 years after transplantation. (A) Preoperative MRI showing cartilage defect at the medial femoral condyle; (B) at 3 years posttransplantation, they observed cartilage regeneration at the defect site; (C) two ROI's were drawn to calculate the change in relaxation rate (ΔR1) in regenerated cartilage and in native cartilage; (D) map by delayed gadolinium-enhanced MRI (dGMRI) of the cartilage shows high glycosaminoglycan (GAG) in the regenerated cartilage. Higher T1 values (arrow 1) reflected an increase of relative GAG content, whereas lower T1 values (arrow 2) are associated with decreased GAG content [25].

6.2. Exam protocol

To calculate proteoglycan content and fixed charge density (FCD) using dGEMRIC, it is required to acquire both precontrast and postcontrast T1 mapping for articular cartilage in addition to a known Gd-DTPA concentration [24]. It has been suggested that native articular cartilage has a relatively constant unenhanced T1 value. So, no need to acquire precontrast images to estimate FCD [26]. However, some authors have shown differences between the precontrast T1 values of ACI repair tissue and articular cartilage [27]. So, it is recommended for the study of cartilage repair tissues to acquire both precontrast and postcontrast T1 measurements when possible [21].

When evaluating cartilage repair tissue using dGEMRIC, we have to know that before contrast injection, repair tissue may show different T1 values compared to normal cartilage. In this case, the postcontrast T1 mapping may not correlate directly with GAG content. So, the solution will be to correlate the difference between precontrast and postcontrast imaging, so called "delta relaxation rate," $\Delta R1 = 1/T1$ precontrast $- 1/T1(Gd)$, which correlates well. Watanabe et al. demonstrated that on study done on 7 patients that the relative ΔR1 or "ΔR1 index" (Δ relaxation rate of repair tissue divided by the Δ relaxation rate of normal hyaline cartilage) correlates with the GAG concentration in ACI repair tissue, using such reference the gas chromatography which is accepted to be the gold standard for the measurement of GAG content in biopsy samples. The limitation of this study was the low number of patients which make statistics low significant [27].

6.3. Spatial resolution

Native articular cartilage and postoperative cartilage repair tissue are relatively thin structures which require very high-resolution images for an accurate assessment. In plane spatial resolution is characterized by the pixel size in both frequency and phase encoding direction. The pixel size is defined as the ratio of the field of view (FOV) over the matrix in both frequency and phase encoding direction whereas the through Plane resolution is characterized by the slice thickness. For accurate assessment of the articular cartilage, it was recommended to use slice thickness less or equal to 2 mm and a pixel size less than 0.3 mm [28] or better less than 0.2 mm [21]. Those recommendations need enough signal to noise ratio (SNR) which can be obtained at higher magnetic field (1.5 T and higher) [21]. This high resolution is recommended to assess fissures which can be developed at the area of peripheral integration as well as the development of proud subchondral bone formation which can be seen after marrow stimulation repair techniques [21].

6.4. Acquisition sequence

In a previous study done on phantom, Trattnig et al. used a 3-D variable flip angle dGEMRIC technique to obtain information related to the long-term development and maturation of grafts in patients after matrix-induced ACI (MACI) surgery. There was a good correlation between variable flip angle technique and standard inversion recovery technique for T1 mapping [29]. Another study also confirmed this correlation *in vivo* [30].

6.5. Clinical dGEMRIC studies on patients with cartilage repair

dGEMRIC has been used to evaluate GAG content in repair tissue after different surgical cartilage repair techniques such as microfracture, ACI, and MACI. Two previous studies reported that MRI non-invasive dGEMRIC technique could be used to monitor the content of GAG after ACI procedure. They suggested that the GAG concentration in repair cartilage after 10 months (or longer) of ACI is comparable with the GAG concentration in the adjacent normal hyaline cartilage [31, 32].

Besides, another study based on MR examinations of 45 patients after cartilage repair surgery using precontrast and dGEMRIC postcontrast, T1 mapping technique revealed a high correlation between T1Gd and ΔR1 in all examinations with R values above −0.8 [33]. From the results, they could assume that both T1Gd and ΔR1 might be useful for evaluation of cartilage repair tissue. Since the T1(Gd) needs only one MRI scan instead of 2 in case of ΔR1, they preferred using T1(Gd) method in order to save time and costs. However, in case they need to compare GAG content of native cartilage and repair tissue within the same patient, the noncontrast T1 values of the native cartilage and repair tissue need to be similar otherwise the comparison may not be valid [21].

In a previous study, a dGEMRIC MRI of cartilage was used to evaluate the quality of the regenerated cartilage at 3 years posttransplantation. The precontrast T1 relaxation time was calculated to evaluate the change in GAG content in the repair-cartilage tissue. The T1 relaxation time was measured in the repair tissue area and the healthy native cartilage. Then, they calculated the relaxation rate R1 as 1/T1 (in 1/second). After, they calculated ΔR1 which represents the change in R1 as the difference of R1 between the precontrast and postcontrast. The ΔR1 represents the concentration of Gd-$DTPA^{2-}$. They defined relative index of ΔR1 as the ratio of ΔR1 in regenerated cartilage divided by ΔR1 in native cartilage. In case of perfect regeneration, this ratio will be equal to 1. The MRI evaluation of five participants after 3 years revealed that the mean relative ΔR1 index was 1.44 which indicated high GAG content of the regenerated cartilage [25, 27].

Trattnig et al. reported that biopsy studies have shown that most of the changes in cartilage implants occur in the early postoperative period. So, in order to assess the maturation of cartilage implants over time, they subdivided patients in 2 groups early postoperative (3–13 months) and late postoperative (19–42 months) groups [29]. In the early postoperative group, the mean ΔR1 (in s^{-1}) for repair tissue was 2.49 (±1.15) versus 1.04 (±0.56) at the intact control site and 1.90 (±0.97) versus with 0.81 (±0.47) in the late postoperative group. The difference in ΔR1 between repair tissue and normal hyaline cartilage in both groups was statistically significant ($P < 0.007$), whereas the difference in ΔR1 of repair tissue and normal hyaline cartilage between the groups was not statistically significant ($P = 0.205$). The mean relative ΔR1 was 2.40 in the early group compared to 2.35 in the late group. They explained this fact by the results of biopsies histological investigations which have shown that MACI may develop hyaline-like, mixed hyaline-fibrous, or fibrous tissue over time [17, 34].

A previous study was conducted on 10 patients treated with microfracture and 10 with MACI. The mean ΔR1 was 1.07 ± 0.34 for microfracture and 0.32 ± 0.20 at the control site, whereas it was 1.90 ± 0.49 for MACI compared to 0.87 ± 0.44 at the control site. Calculated relative ΔR1 was 3.39 for microfracture and 2.18 for MACI and the difference between the cartilage repair groups was statistically significant [35]. The histology and biochemistry analyze showed that the repair tissue formed by microfracture contained less PGs and an abnormal distribution of collagen compared with normal cartilage which may explain the poor mechanical properties often exhibited by repair tissue. The T1 mapping showed a significantly higher relative ΔR1 of the repair tissue after microfracture when compared after MACI, suggesting a lower GAG content after microfracture [36, 37].

In one study, Fibrocartilage formed after microfracture, evaluated using dGEMRIC, demonstrated a greater difference between precontrast and postcontrast T1 relaxation time compared with repair tissue formed after MACI. As dGEMRIC reflects the glycosaminoglycan content, they deduced from the results that glycosaminoglycan content in fibrocartilage were lower compared to other types of cartilage repair tissue [35].

Another study conducted on nine patients (average age, 21.2 years) reported that relative ΔR1 index was 1.32 after 1 year post-ACI for focal chondral defects. In nine patients (average age, 43.2 years) postosteochondral allograft transplantation, relative ΔR1 index were 1.13 at the first year and 1.55 at the second year [38].

7. T2 mapping

During the relaxation process of MRI experiment and due to the variations of the local magnetic field, the individual magnetic moments gradually lose their phase coherence, which leads to a decrease of the net magnetization vector. This decrease of the signal is called spin-spin relaxation and noted T2 relaxation. The calculation of T2 mapping is obtained usually with a spin echo sequence using different echo times (**Figure 2**). From the signals measured with different TE's, we draw the T2 decay curve where T2 correspond to the time spent by the transverse relaxation magnetization to reach 37% from its initial value. T2 maps are usually obtained by using a pixelwise, monoexponential, non-negative least-squares fit analysis (**Figure 3**).

The T2 relaxation time is affected by the speed the spins lose phase coherence during relaxation. The presence of free water molecules in knee cartilage will slow the decay of the transverse magnetization which will make from the T2 mapping a common tool to measure the water content in the cartilage [39].

T2 value is also affected mostly by collagen network structure of cartilage [19]. It depends on both water [16, 17] and collagen content [18]. The concentration of collagen and proteoglycans is responsible for the water movements in the extracellular matrix and the appearance of the cartilage in T2-weighted images. Quantitative T2 MR mapping of articular cartilage is a non-invasive imaging technique that has the potential to characterize hyaline articular cartilage and repair tissue. The T2 relaxation time has been significantly correlated with collagen orientation in cartilage repair models using either polarized light microscopy or Fourier transform infrared imaging spectroscopy [40–42] where as it showed a poor correlation with collagen content in several repair models [42, 43].

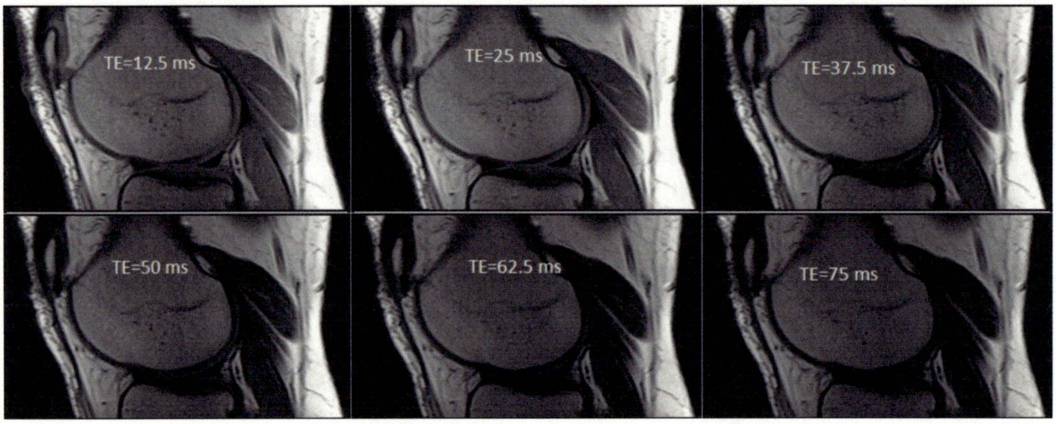

Figure 2. Images acquired using multi echo spin echo (MESE) sequence with different TE's in the range of 12.5–75 ms.

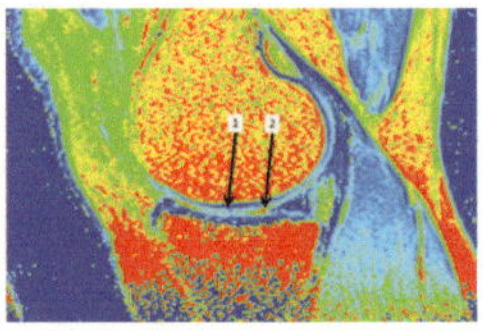

Figure 3. T2 mapping image calculated using a MESE sequence. Arrow 2 indicates higher T2 whereas arrow 1 indicates lower T2 values.

7.1. Spatial variation of T2 values

The T2 relaxation time is affected by the organization of the extracellular matrix of native articular cartilage [6]. In native hyaline cartilage, the T2 relaxation times is varying over depth when going from deepest layers to superficial layers with shorter T2 values in the deeper, radial zone, where the collagen is highly ordered and the collagen fiber matrix has a preferred orientation perpendicular to the cartilage surface, and longer values in the transitional zone because of less organization of the collagen where the collagen fiber matrix has an oblique orientation. The superficial zone may not be visualized on morphological imaging and quantitative T2 mapping because it is too thin [44].

When doing quantitative MR T2 mapping in the knee articular cartilage to compare different cartilage repair surgeries, we can either evaluate mean global T2 value throughout the thickness of the repair or a zonal assessment in the deep versus the superficial half of the repair tissue. Cartilage repair tissue with a lack of zonal organization of collagen would not be expected to demonstrate a similar of T2 values from the deep to superficial aspects of the tissue compared to normal cartilage. Alteration in this orderly transition in T2 values within cartilage has been shown to correlate to changes in water content and changes in collagen structure and organization associated with hyaline articular cartilage degradation [45].

7.2. T2 mapping sequences

The common point between the sequences used for T2 mapping calculation is the acquisition of multiechoes to describe the T2 decay curve and to allow the calculation of T2 value. Among those sequences we found spin echo single echo sequence (SESE), multiecho spin echo sequence (MESE), dual echo steady state sequence (DESS) and turbo gradient spin echo (TGSE).

7.2.1. Spin echo single echo sequence (SESE)

The SESE Sequence uses two RF pulses, 90 and 180° pulses. The 90° pulse will tilt the longitudinal magnetization vector M0 to the measurement plane which is the transverse plane. The spins start dephasing. Then we apply the 180° pulse to rephase spins. At a certain time called echo time (TE) when the spins are totally rephased we measure the signal. Then we repeat the pulse sequence many times as much as the phase encoding matrix. The Time which separates two consecutive 90° pulses is called repetition time (TR). In this sequence, we measure a single echo in each repetition time (TR). To calculate the T2 relaxation time, we need to repeat the sequence many times in order to collect different TE's. The main advantage of this type of sequence is that it is not contaminated by the stimulated echo. Also, this sequence, by the use of 180° refocusing pulses is less sensitive to artifacts in case of postoperative imaging [21]. The disadvantages are that the exam duration will be longer adding to that the risk of patient's movement.

7.2.2. Multiecho spin echo sequence (MESE)

The MESE sequence uses the same preparation radio frequency (RF) pulses as the SESE. The difference is that in SESE sequence we measure only one echo in a TR where as in the MESE sequence, we can measure many echoes. The biggest advantage is that we measure all the TE's on one scan which will save time with less movement artifact. In addition, this sequence gives the possibility to measure the T2 using the inline calculation method. The only disadvantage is the presence of the stimulated echo which can be reduced by the elimination of the first echo from the calculation.

7.2.3. Dual echo steady state sequence (DESS)

T2 can be calculated using dual echo steady state sequence (DESS) which demonstrated to provide results as comparable with the standard multiecho spin echo T2 [2]. In both 2D fat suppressed turbo spin echo proton density and 3D DESS sequence, hyaline cartilage has intermediate signal and synovial fluid has high signal. 3D DESS has the advantage to use thinner slices which make this sequence to me sensitive to detect smaller cartilage defects better than the 2D sequence.

7.2.4. Turbo gradient spin echo

This sequence combines a gradient echo and a spin echo imaging. It generates additional gradient echo before and after each spin echo. The spin echo gives the T2 contrast and the gradient echo determines the image resolution. The main advantage of the sequence is that it is fast and provides high resolution images. The TGSE sequence combines the TSE and echo-planar imaging method. It provides T2-weighted images. There are technical differences between TSE and TGSE sequences that could make the contrast and signal-to-noise ratio potentially different.

7.3. Technical aspects

When optimizing a T2 mapping acquisition protocol, we need to take into account many technical considerations.

7.3.1. Repetition time (TR)

To minimize the T1 contribution in the image contrast of T2 images, it is recommended to use longer TR value compared to the T1 value of the articular cartilage. A TR of 1500 ms or longer is preferred.

7.3.2. Echo time (TE)

Due to the shorter value of the T2 relaxation time of the knee articular cartilage, short TE and short echo spacing (ES) in case of multiecho sequences are required to accurately characterize the T2 decay curve. Since the expected T2 values of articular cartilage are in the range between 20 and 70 ms, we recommend using many echoes for better curve fitting. The greater the number of data sets, that is the number of TE values, the greater the accuracy of the T2 measurements but without using a higher TE which is susceptible to greater noises and errors.

7.3.3. Stimulated echo

It is important to know that multi-slice multiecho spin echo sequences (MS MESE) uses a slice-selective refocusing pulses. In case of bad calibration or inhomogeneities of the radio frequency pulse, slice-selective refocusing pulses do not result in rectangular slice profiles causing stimulated echo contributions to the measured signal. The T2 relaxation time based on multiecho sequence is subject of measurement errors because of the stimulated echo which may increase artificially the T2 value [46]. This error may be avoided by ignoring the first echo when using a multiecho sequence or by using single echo acquisitions instead of multiecho acquisition.

7.3.4. Bandwidth (BW)

To reduce the chemical shift artifact between water and fat in the cartilage, we advise to use a higher bandwidth of ~217 Hz/pixel corresponding to a chemical shift of 1 pixel on 1.5 T system and 0.5 pixel at 3 T.

7.3.5. Magnetic field

MR morphology imaging of cartilage repair tissue has significantly improved in recent years by the use of high-field MR systems like 3 T, the use of higher gradient strengths and the dedicated coils. This improvement increased the signal to noise ratio (SNR) which allows high-resolution imaging of cartilage within reasonable scan time [6]. To further decrease the scan time while maintaining high-resolution, most of the new systems used a dedicated multi-elements coil which enables the use of parallel acquisition techniques with high acceleration factor [21].

High-field MRI system also allows the use of 3-D acquisition sequences with the advantage of isotropic high resolution where dimensions are equal in all 3 axes (frequency, phase and slice) while maintaining high SNR and high contrast-to-noise ratios (CNR). This permits multi-planar reconstruction (MPR) in any plane with the same resolution. Biochemical imaging techniques, such as sodium MR imaging, which is limited by low signal-to-noise ratio at standard clinical field strength can be used at ultra–high magnetic field [12].

Care must be taken when performing T2 mapping and interpreting the results since T2 may depend on Bo, with shorter T2 values found at higher field strengths.

7.3.6. *Magic angle effect*

One of the disadvantages of T2 relaxation time mapping is its susceptibility to the magic angle effect, in which T2 values may be artificially elevated in certain regions according to the orientation of cartilage in relation to the main magnetic field [5]. The magic angle effect may complicate evaluation of curved articular surfaces, such as the femoral condyle [47], and should not be misinterpreted as degeneration. However, a recent report has found that OA may affect T2 values to a greater degree than the magic angle effect [48]. This finding may enable utilization of magic angle T2 mapping data with the understanding that only regions of interest from similar anatomic locations may be compared. However, the magic angle effect should not impact results tracking changes over time or between study populations as long as the subjects are positioned in the same manner in the magnet [49].

7.3.7. *Exam timing*

Significant differences between cartilage T2 values were obtained at the beginning and at the end of the MRI examination resulting from the different states of unloading of the knee in the course of the MRI examination due to the supine position of the patient. Therefore, the time point of T2 acquisition has to be considered in the MRI protocol. Apprich et al. recommended to measure T2 after unloading, i.e., at the end of the MRI examination [46].

7.3.8. *Question to be answered*

The following questions have to be answered in case of cartilage repair follow-up: (1) are there different *T2* relaxation times between repair tissues and adjacent native cartilage? (2) Are these differences reduced over time? (3) Is there a difference between a global assessment and line profile assessment? [6].

7.4. Clinical application of T2 mapping in cartilage repair surgery

In a previous study conducted by Welsch et al., they calculated the mean and the zonal T2 values within the repair tissue and hyaline native cartilage on twenty patients who underwent MFX or MACT (10 in each group) with minimum 2-year follow-up. They compared cartilage T2 values after microfracture therapy (MFX) and matrix-associated autologous chondrocyte transplantation (MACT) repair procedures. They reported that in normal native hyaline cartilage, T2 values showed similar values for all patients with a significant increase of T2 values from deep to superficial zones ($P < 0.05$). In cartilage repair areas after MFX, global mean T2 was significantly decreased ($P < 0.05$), whereas cartilage repair areas after MACT showed no decrease of mean T2 ($P \geq 0.05$). For zonal variation, repair tissue after MFX showed no significant trend between different depths ($P \geq 0.05$), in contrast to repair tissue after MACT which showed a significant increase of T2 values from deep to superficial zones ($P < 0.05$) [50] (**Figure 4**).

In another study, Welsch et al. compared T2 mapping of 17 patients who underwent MACT over the patella versus 17 patients who underwent MACT in the medial femoral condyle. They reported an increase of T2 values over the condyle compared to the patella repair tissue. They conclude that differential maturation of the repair tissue depends on its environment [51].

Welsch and colleagues reported in another study that T2 mapping can be used to distinguish between MACI performed using a collagen-based scaffold and a hyaluronan-based scaffold (higher T2 values in collagen-based scaffolds) [52].

Quantitative T2 mapping has been used to assess the interface between transplanted and native cartilage. A clinical study of patellar autologous osteochondral transplantation reported progressive T2 increase at the offset of the tidemark that occurred between the thicker native cartilage over the patella and the thinner cartilage over the autologous plug [53].

A study of T2 mapping performed in 53 sites reported a perfect agreement between organized T2 and histologic findings of hyaline cartilage and between disorganized T2 and histologic findings of fibrous reparative tissue (k = 1.0). Mean T2 values were 53.3, 58.6, and 54.9 ms at the deep, middle, and superficial cartilage, respectively, at reparative fibrous tissue, whereas T2 mean values were 40.7, 53.6, and 61.6 ms at hyaline cartilage. A significant increase of T2 values (from deep to superficial) was found in hyaline cartilage ($P < 0.01$). Fibrous tissue sites showed no significant change with depth ($P > 0.59$) [45].

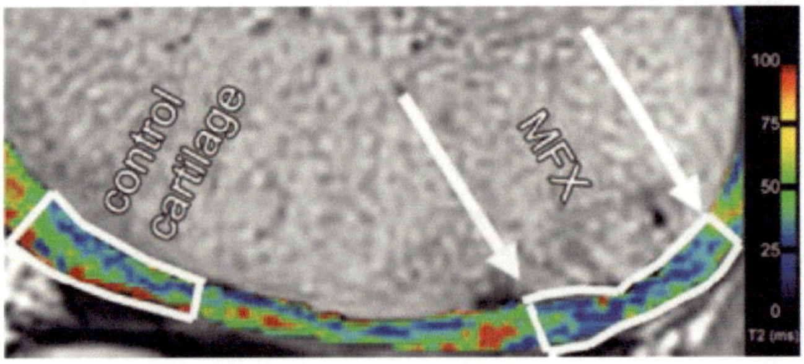

Figure 4. Enlarged section of sagittal cartilage T2 map. ROI of cartilage repair (between two arrows) shows no zonal variation and low T2 values, whereas control cartilage shows visible zonal variation from deep to superficial areas, with higher T2 values in superficial area [50].

Two previous studies evaluated the status of reparative fibrocartilage induced by microfracture using T2 mapping. They reported that spatial variation of T2 values in fibrocartilage and native hyaline cartilage were not the same (hyaline cartilage is characterized by higher T2 values near the articular surface and lower T2 values near subchondral bone) [41, 50]. Also, the overall global T2 value for fibrocartilage repair tissue was lower compared to native hyaline cartilage [50].

MACI has been studied using T2 mapping. The results showed similar spatial variation in the T2 values of repair cartilage like seen in native hyaline cartilage (although the increase in mean T2 values from deep to superficial layers of cartilage is less pronounced) [41].

T2 mapping of patients after MACT surgery at different postoperative intervals Quantitative *T*2 mapping was performed in 15 patients after MACT surgery at different postoperative intervals. With respect to the postoperative time interval, patients were subdivided into two groups: group I, 3–13 months (6 patients); group II, 19–42 months (9 patients). In group I, the mean global *T*2 values in cartilage repair tissue was 65.8 ± 16.6 compared with 50.0 ± 7.0 for native cartilage; this difference was statistically significant ($P = 0.013$). In group II, the mean *T*2 values of repair tissue were 56.5 ± 12.0 compared with 57.7 ± 9.2 for native cartilage. These differences were not statistically significant ($P = 0.784$). Results showed significantly higher *T*2 values, in cartilage repair tissue, in the early stage (3–13 months) compared with native hyaline cartilage. Over time, there was a decrease in repair tissue *T*2 values which became similar to native healthy cartilage [6]. This finding is in agreement with a study by Kurkijarvi et al. [54] who, in 1.5 T, reported *T*2 values in the repair tissue and normal hyaline cartilage with 60 ± 10 ms and 50 ± 7 ms, respectively, in 10 patients 10–15 months after ACI surgery.

Domayer et al. introduced a T2 index defined as the ratio of the mean global repair tissue T2 divided the mean global normal cartilage expressed as a percentage. They reported that this T2 index correlated with clinical measurements [55].

8. T2* mapping

In addition to the phase shift of the individual spins, there is also the additional phase shift caused by field inhomogeneities that increase the phase shift of the spins and thus accelerates the decay. The total relaxation time (T2*) is a consequence of these terms.

8.1. T2* mapping principle

The physical difference between T2* and T2 is that magnetic gradients, and not a 180° RF pulse, are used to rephase the spins at a user defined TE. T2* and T2 values are related by the equation (Eq. (1)):

$$\frac{1}{T2^*} = \frac{1}{T2} + \gamma\, \Delta B0 \tag{1}$$

Where γ is the gyromagnetic ratio of the observed nucleus and $\Delta B0$ is the magnetic field inhomogeneity. If we assume that the applied static magnetic field B0 is uniform then $\gamma\, \Delta B0$

is only influenced by local magnetic susceptibility fields. In the case of knee articular cartilage, this susceptibility will be present at the cartilage bone interface or within the cartilage microstructure.

T2* mapping is similar to T2 mapping [56]: multiple echo images at a single slice location are generated, and a mono- or bi-exponential decay equation [57] (**Figure 5**) is used to fit the signal intensity to the corresponding echo time data. The difference between T2 and T2* mapping is that T2 mapping is calculated using a spin echo sequence however T2* mapping is obtained using a gradient echo sequence (**Figure 6**). T2* mapping has the advantage of shorter scan time compared to T2 mapping. Also, with T2*, we can acquire shorter TE compared to T2 which is very important for short T2 components. In addition, with T2* mapping using 3D gradient echo sequence, we have the possibility of isotropic three-dimensional reconstruction, which seems to offer a potential alternative and reliable results in cartilage imaging [58].

8.2. Clinical application of T2* mapping in cartilage repair surgery

Goetz H. and al performed MRI examinations on 30 patients after MACT at a follow-up period of 28.1 ± 18.8 months. T2* values are given in milliseconds (ms). In healthy control cartilage, T2* mean value of all patients was 30.9 ± 6.6 with a significant increase of T2* values from deep (27.9 ± 7.2) to superficial (33.9 ± 6.9) cartilage aspects. The cartilage repair tissue after MACT showed a mean (full-thickness) T2* value of 24.5 ± 8.1 with a significant increase

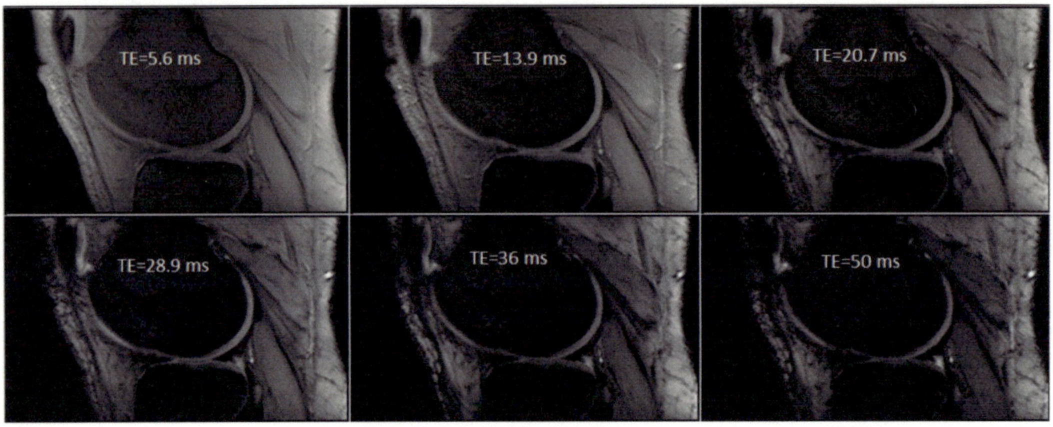

Figure 5. Images acquired using multiecho gradient echo (MEGE) sequence with different TE's in the range of 5.1–50 ms.

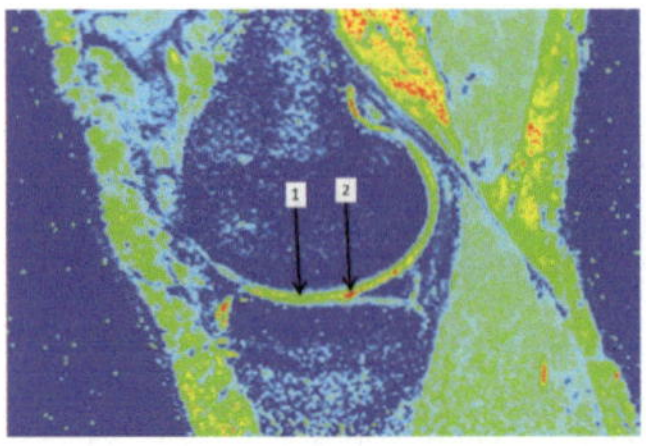

Figure 6. T2* mapping image calculated using a MEGE sequence. Arrow 2 indicates higher T2* whereas arrow 1 indicates lower T2* values.

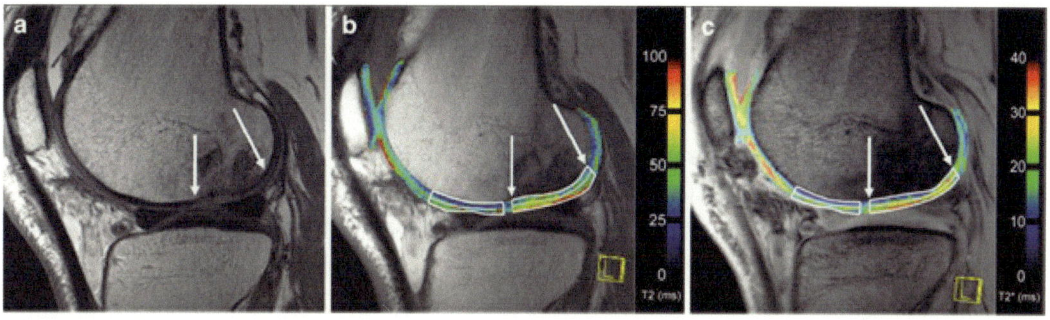

Figure 7. Depiction of cartilage in a patient 6 months after MACT of the lateral femoral condyle. Morphological PD-TSE sequence (a), matched quantitative T2 (b), and T2* (c) maps. Arrows mark the area of cartilage repair. ROIs, considering a possible zonal variation, provide information on the mean (full-thickness) as well as the deep and superficial aspect of control cartilage (left) and cartilage repair tissue (right, arrows). Zonal stratification is visible for both T2 and T2* images in most parts of the cartilage. A possible "magic angle" effect is visible within the trochlea. Higher T2/T2* values are apparent in the cartilage repair tissue, compared with the adjacent cartilage [59].

from deep (21.6 ± 7.3) to superficial (27.5 ± 9.4) ($P < 0.001$). When comparing T2* values of the healthy control cartilage with those of the cartilage repair tissue, the mean T2* values and the T2* values in deep and superficial cartilages were significantly lower in the cartilage repair tissue ($P < 0.001$) [59] (**Figure 7**).

The comparison of the mean (full-thickness) T2* values over different postoperative intervals revealed a stability of T2* values over time with T2* value of 31.4 ± 6.2 for the short-term interval, 31.0 ± 6.7 for the mid-term interval and 30.4 ± 7.0 for the long-term interval. However, the cartilage repair tissue showed significantly higher T2* values at the short-term follow-up (31.0 ± 8.1) than at the mid-term follow-up (20.7 ± 6.1) ($P < 0.001$), and stable values between the mid-term and long-term (22.2 ± 6.0) follow-up ($P = 0.232$). The difference between the short-term and long-term follow-up was also significant ($P < 0.001$) [59].

The comparison of mean (full-thickness) T2* values for healthy control cartilage and cartilage repair tissue at the different postoperative follow-up time points revealed comparable values at the short-term follow-up (0.793), significantly lower mean (full-thickness) T2* values in cartilage repair tissue compared to healthy control cartilage for the mid-term ($P < 0.001$) and long-term ($P < 0.001$) postoperative intervals [59].

Goetz H. Welsch and al reported that mean T2* values (ms) were lower at 7 T (18.3 ± 4.9) compared with 3 T (22.2 ± 4.3). Regarding zonal variation, T2* relaxation times (ms) were significantly lower at 7 T (deep: 15.5 ± 3.7; superficial: 21.0 ± 4.5) ($P < 0.001$) compared with 3 T (mean: deep: 17.6 ± 3.7; superficial: 26.9 ± 5.4) [60].

9. Conclusion

The validation of cartilage repair techniques needs short, medium and long term follow-up. The follow-up periods remain a problem for cartilage repair because of the slow progression of cartilage degeneration over time. Choosing the best technique that addresses the individual defect is a challenge for the orthopedic surgeon. T2 mapping could provide information about collagen matrix concentration and organization, whereas dGEMRIC is sensitive to proteoglycan content. T2* mapping has the advantage of shorter scan time with the possibility to acquire shorter TE compared to T2 which is very important for shorter T2 components. The modifications of the extracellular matrix, such as loss PG, may be reflected by the change of the T1ρ values. Each MRI parameter can characterize certain features of the articular cartilage properties. All together may provide complementary information's about cartilage repair tissue properties.

Author details

Mars Mokhtar

Address all correspondence to: mokhtar.mars-mms@topnet.tn

Biophysics and Medical Technologies Laboratory, Institut Supérieur des Technologies Médicales de Tunis, Université de Tunis El Manar, Tunis, Tunisia

References

[1] Baum T, Joseph GB, Karampinos DC, Jungmann PM, Link TM, Bauer JS. Cartilage and meniscal T2 relaxation time as non-invasive biomarker for knee osteoarthritis and cartilage repair procedures. Osteoarthritis and Cartilage. 2013;**21**:1474-1484

[2] Ahlawat S, Padua A, Huisman TAGM, Carrino JA. 3T MR Imaging of the Pediatric Cartilage Using 3D Dual Echo Steady State (DESS). MAGNETOM Flash I 2/2014 I www.siemens.com/magnetom-world, Germany

[3] Lusse S, Claassen H, Gehrke T, Hassenpflug J, Schunke M, Heller M, et al. Evaluation of water content by spatially resolved transverse relaxation times of human articular cartilage. Magnetic Resonance Imaging. 2000;**18**(4):423-430

[4] Nieminen MT, Toyras J, Rieppo J, Hakumaki JM, Silvennoinen J, Helminen HJ, et al. Quantitative MR microscopy of enzymatically degraded articular cartilage. Magnetic Resonance in Medicine. 2000;**43**:676-681

[5] Xia Y. Magic-angle effect in magnetic resonance imaging of articular cartilage: A review. Investigative Radiology. 2000;**35**:602-621

[6] Trattnig S, Mamisch TC, Welsch GH, Glaser C, Szomolanyi P, Gebetsroither S, et al. Quantitative T2 mapping of matrix-associated autologous chondrocyte transplantation at 3 Tesla: An in vivo cross-sectional study. Investigative Radiology. June 2007;**42**(6):442-448

[7] Choi YS, Potter HG, Chun TJ. MR imaging of cartilage repair in the knee and ankle. RadioGraphics. 2008;**28**:1043-1059

[8] Smith GD, Knutsen G, Richardson JB. A clinical review of cartilage repair techniques. The Journal of Bone and Joint Surgery. April 2005;**87-B**(4):445-449

[9] Goyal D, Keyhani S, Lee EH, Hui JHP. Evidence based status of microfracture technique: A systematic review of level I and II studies. Arthroscopy. 2013;**29**:1579-1588

[10] Mithoefer K, McAdams T, Williams RJ, Kreuz PC, Mandelbaum BR, et al. Clinical efficacy of the microfracture technique for articular cartilage repair in the knee: An evidence-based systematic analysis. The American Journal of Sports Medicine. 2009;**37**:2053-2063

[11] Lee JJ, Lee SJ, Lee TJ, Yoon TH, Choi CH. Results of microfracture in the osteoarthritic knee with focal full-thickness articular cartilage defects and concomitant medial meniscal tears. Knee Surgery & Related Research. 2013;**25**:71-76

[12] Chang G, Sherman O, Madelin G, Recht M, Regatte R. MR imaging assessment of articular cartilage repair procedures. Magnetic Resonance Imaging Clinics of North America. 2011 May;**19**(2):323-337

[13] Choi YS, Potter HG, Chun TJ. MR imaging of cartilage repair in the knee and ankle. Radiographics. 2008 Jul-Aug;**28**(4):1043-1059 Review

[14] Marlovits S, Striessnig G, Resinger CT, Aldrian SM, Vecsei V, Imhof H, et al. Definition of pertinent parameters for the evaluation of articular cartilage repair tissue with high-resolution magnetic resonance imaging. European Journal of Radiology. 2004;**52**:310-319

[15] Marlovits S, Singer P, Zeller P, Mandl I, Haller J, Trattnig S. Magnetic resonance observation of cartilage repair tissue (MOCART) for the evaluation of autologous chondrocyte transplantation: Interobserver variability and correlation to clinical outcome after 2 years. European Journal of Radiology. 2006;**57**:16-23

[16] Brown WE, Potter HG, Marx RG, Wickiewicz TL, Warren RF. Magnetic resonance imaging appearance of articular cartilage repair in the knee. Clinical Orthopaedics and Related Research. 2004;**422**:214-223

[17] Tins BJ, McCall IW, Takahashi T, Cassar-Pullicino V, Roberts S, Ashton B, et al. Autologous chondrocyte implantation in knee joint: MR imaging and histologic features at 1-year follow-up. Radiology. 2005;**234**(2):501-508

[18] Trattnig S, Ba-Ssalamah A, Pinker K, Plank C, Vecsei V, Marlovits S. Matrix-based autologous chondrocyte implantation for cartilage repair: Noninvasive monitoring by high-resolution magnetic resonance imaging. Magnetic Resonance Imaging. 2005;**23**:779-787

[19] Redman SN, Oldfield SF, Archer CW. Current strategies for articular cartilage repair. European Cells and Materials. 2005;**9**:23-32

[20] Mow VC, Zhu W, Ratcliffe A. Structure and function of articular cartilage and meniscus. In: Mow VC, Hayes WE, editors. Basic Orthopaedic Biomechanics. New York: Raven Press; 1991. p. 143-198

[21] Trattnig S, Winalski CS, Marlovits S, Jurvelin JS, Welsch GH, Potter HG. Magnetic resonance imaging of cartilage repair: A review. Cartilage. 2011;**2**:5 originally published online 18 April 2010

[22] Trattnig S, Millington SA, Szomolanyi P, Marlovits S. MR imaging of osteochondral grafts and autologous chondrocyte implantation. European Radiology. 2007;**17**(1):103-118

[23] Deoni SC, Rutt BK, Peters TM. Rapid combined T1 and T2 mapping using gradient recalled acquisition in the steady state. Magnetic Resonance in Medicine. 2003;**49**:515-526

[24] Fram EK, Herfkens RJ, Johnson GA, Glover GH, Karis JP, Shimakawa A, et al. Rapid calculation of T1 using variable flip angle gradient refocused imaging. Magnetic Resonance Imaging. 1987;**5**:201-208

[25] Park Y-B, Ha C-W, Lee C-H, Yoon YC, Park Y-G. Cartilage regeneration in osteoarthritic patients by a composite of allogeneic umbilical cord blood-derived mesenchymal stem cells and hyaluronate hydrogel: Results from a clinical trial for safety and proof-of-concept with 7 years of extended follow-up. Stem Cells Translational Medicine. 2016;**5**:1-9

[26] Burstein D, Velyvis J, Scott KT, Stock KW, Kim YJ, Jaramillo D, et al. Protocol issues for delayed Gd(DTPA)(2-)-enhanced MRI: (dGEMRIC) for clinical evaluation of articular cartilage. Magnetic Resonance in Medicine. 2001;**45**(1):36-41

[27] Watanabe A, Wada Y, Obata T, Ueda T, Tamura M, Ikehira H, et al. Delayed gadolinium-enhanced MR to determine glycosaminoglycan concentration in reparative cartilage after autologous chondrocyte implantation: Preliminary results. Radiology. 2006; **239**(1):201-208

[28] Rubenstein JD, Li JG, Majumdar S, Henkelman RM. Image resolution and signal-to-noise ratio requirements for MR imaging of degenerative cartilage. AJR. American Journal of Roentgenology. 1997;**169**(4):1089-1096

[29] Trattnig S, Marlovits S, Gebetsroither S, Szomolanyi P, Welsch GH, Salomonowitz E, et al. Three-dimensional delayed gadolinium-enhanced MRI of cartilage (dGEMRIC) for in vivo evaluation of reparative cartilage after matrix-associated autologous chondrocyte transplantation at 3.0T: Preliminary results. Journal of Magnetic Resonance Imaging. 2007;**26**(4):974-982

[30] Mamisch TC, Dudda M, Hughes T, Burstein D, Kim YJ. Comparison of delayed gadolinium enhanced MRI of cartilage (dGEMRIC) using inversion recovery and fast T1 mapping sequences. Magnetic Resonance in Medicine. 2008;**60**(4):768-773

[31] Kurkijarvi JE, Mattila L, Ojala RO, Vasara AI, Jurvelin JS, Kiviranta I, et al. Evaluation of cartilage repair in the distal femur after autologous chondrocyte transplantation using T-2 relaxation time and dGEMRIC. Osteoarthritis and Cartilage. 2007;**15**(4):372-378

[32] Gillis A, Bashir A, McKeon B, Scheller A, Gray ML, Burstein D. Magnetic resonance imaging of relative glycosaminoglycan distribution in patients with autologous chondrocyte transplants. Investigative Radiology. 2001;**36**(12):743-748

[33] Trattnig S, Burstein D, Szomolanyi P, Pinker K, Welsch GH, Mamisch TC. T1(Gd) gives comparable information as delta T1 relaxation rate in dGEMRIC evaluation of cartilage repair tissue. Investigative Radiology. 2009;**44**(9):598-602

[34] Nehrer S, Minas T. Treatment of articular cartilage defects. Investigative Radiology. 2000;**35**(10):639-646

[35] Trattnig S, Mamisch TC, Pinker K, Domayer S, Szomolanyi P, Marlovits S, et al. Differentiating normal hyaline cartilage from post-surgical repair tissue using fast gradient echo imaging in delayed gadolinium-enhanced MRI (dGEMRIC) at 3 Tesla. European Radiology. 2008;**18**(6):1251-1259

[36] Minas T, Nehrer S. Current concepts in the treatment of articular cartilage defects. Orthopedics. 1997;**20**(6):525-538

[37] Ghivizzani SC, Oligino TJ, Robbins PD, Evans CH. Cartilage injury and repair. Physical Medicine and Rehabilitation Clinics of North America. 2000;**11**(2):289-307 vi

[38] Brown DS, Durkan MG, Foss EW, Szumowski J, Crawford DC. Temporal in vivo assessment of fresh osteochondral allograft transplants to the distal aspect of the femur by dGEMRIC (delayed gadolinium-enhanced MRI of cartilage) and zonal T2 mapping MRI. The Journal of Bone and Joint Surgery. American Volume. 2014;**96**:564-572

[39] Matzat SJ, van Tiel J, Gold GE, Oei EHG. Quantitative MRI techniques of cartilage composition. Quantitative Imaging in Medicine and Surgery. 2013;**3**(3):162-174

[40] Kelly BT, Potter HG, Deng XH, Pearle AD, Turner AS, Warren RF, et al. Meniscal allograft transplantation in the sheep knee: Evaluation of chondroprotective effects. The American Journal of Sports Medicine. 2006;**34**(9):1464-1477

[41] White LM, Sussman MS, Hurtig M, Probyn L, Tomlinson G, Kandel R. Cartilage T2 assessment: Differentiation of normal hyaline cartilage and reparative tissue after arthroscopic cartilage repair in equine subjects. Radiology. 2006;**241**(2):407-414

[42] Kim M, Foo L, Lyman S, Ryaby JT, Grande DA, Potter HG, et al. Evaluation of early osteochondral defect repair in a rabbit model utilizing Fourier transform infrared imaging spectroscopy (FT-IRIS), magnetic resonance imaging (MRI) and quantitative T2 mapping. Tissue Eng Part C Methods. 2010 Jun;**16**(3):355-364

[43] Watanabe A, Boesch C, Anderson SE, Brehm W, Mainil Varlet P. Ability of dGEMRIC and T2 mapping to evaluate cartilage repair after microfracture: A goat study. Osteoarthritis and Cartilage. 2009;**17**(10):1341-1349

[44] Potter HG, Foo LF. Magnetic resonance imaging of articular cartilage: Trauma, degeneration, and repair. The American Journal of Sports Medicine. 2006;**34**(4):661-677

[45] White LM, Sussman MS, Hurtig M, Probyn L, Tomlinson G, Kandel R. Cartilage T2 assessment: Differentiation of normal hyaline cartilage and reparative tissue after arthroscopic cartilage repair in equine subjects. Radiology. November 2006;**241**(2):407-414

[46] Maier CF, Tan SG, Hariharan H, Potter HG. T2 quantitation of articular cartilage at 1.5 T. Journal of Magnetic Resonance Imaging. 2003;**17**:358-364

[47] Mosher TJ, Smith H, Dardzinski BJ, Schmithorst VJ, Smith MB. MR imaging and T2 mapping of femoral cartilage: In vivo determination of the magic angle effect. American Journal of Roentgenology. 2001;**177**:665-669

[48] Wang L, Regatte RR. Investigation of regional influence of magic-angle effect on t2 in human articular cartilage with osteoarthritis at 3 T. Academic Radiology. 2015; **22**:87-92

[49] Matzat SJ, McWalter EJ, Kogan F, Chen W, Gold GE. T2 relaxation time quantitation differs between pulse sequences in articular cartilage. Journal of Magnetic Resonance Imaging. 2015;**42**:105-113

[50] Welsch GH, Mamisch TC, Domayer SE, Dorotka R, Kutscha-Lissberg F, Marlovits S, et al. Cartilage T2 assessment at 3-T MR imaging: In vivo differentiation of normal hyaline cartilage from reparative tissue after two cartilage repair procedures—Initial experience. Radiology. April 2008;**247**(1):154-161

[51] Welsch GH, Mamisch TC, Quirbach S, Zak L, Marlovits S, Trattnig S. Evaluation and comparison of cartilage repair tissue of the patella and medial femoral condyle by using morphological MRI and biochemical zonal T2 mapping. European Radiology. 2009;**19**(5):1253-1262

[52] Welsch GH, Mamisch TC, Zak L, Blanke M, Olk A, Marlovits S, et al. Evaluation of cartilage repair tissue after matrix-associated autologous chondrocyte transplantation using a hyaluronic-based or a collagen-based scaffold with morphological MOCART scoring and biochemical T2 mapping: Preliminary results. The American Journal of Sports Medicine. 2010;**38**(5):934-942

[53] Nho SJ, Foo LF, Green DM, Shindle MK, Warren RF, Wickiewicz TL, et al. Magnetic resonance imaging and clinical evaluation of patellar resurfacing with press-fit osteochondral autograft plugs. The American Journal of Sports Medicine. 2008;**36**(6):1101-1109

[54] Kurkijärvi JE, Nissi M, Ojala RO, Vasara AI, Jurvelin JS, Kiviranta I, et al. In vivo T2 mapping and dGEMRIC of human articular cartilage repair after autologous chondrocyte transplantation. ISMRM 13th Scientific Meeting & Exhibition in Miami Beach, Florida, USA. 2005;**13**:481

[55] Domayer SE, Kutscha-Lissberg F, Welsch G, Dorotka R, Nehrer S, Gabler C, et al. T2 mapping in the knee after microfracture at 3.0 T: Correlation of global T2 values and clinical outcome. Preliminary results. Osteoarthritis and Cartilage. 2008;**16**(8):903-908

[56] Eagle S, Potter HG, Koff MF. Morphologic and quantitative magnetic resonance imaging of knee articular cartilage for the assessment of post-traumatic osteoarthritis. Journal of Orthopaedic Research. March 2017;**35**(3):412-423

[57] Bittersohl B, Miese FR, Hosalkar HS, Herten M, Antoch G, Krauspe R, et al. T2* mapping of hip joint cartilage in various histological grades of degeneration. Osteoarthritis and Cartilage. 2012;**20**:653-660

[58] Murphy BJ. Evaluation of grades 3 and 4 chondromalacia of the knee using T2*-weighted 3D gradient-echo articular cartilage imaging. Skeletal Radiology. 2001;**30**:305-311

[59] Welsch GH, Trattnig S, Hughes T, Quirbach S, Olk A, Blanke M, et al. T2 and T2* mapping in patients after matrix-associated autologous chondrocyte transplantation: Initial results on clinical use with 3.0-Tesla MRI. European Radiology. 2010;**20**:1515-1523

[60] Welsch GH, Apprich S, Zbyn S, Mamisch TC, Mlynarik V, Scheffler K, et al. Biochemical (T2, T2* and magnetisation transfer ratio) MRI of knee cartilage: Feasibility at ultra-high field (7T) compared with high field (3T) strength. European Radiology. 2011;**21**:1136-1143

Section 3

Head and Neck

Chapter 11

Applied Basic Science of the Auricular Cartilage

Mohamed Khamis Tolba Mahmoud Abdalla

Additional information is available at the end of the chapter

http://dx.doi.org/10.5772/intechopen.72479

Abstract

Cartilage is an essential component of human body, and it is present in any region of the body. Auricular cartilages play an essential role in esthetic aspect and shape of the face. Therefore, comprehensive understanding of the applied basic science of the cartilage of the ear is essential to understand the pathophysiology of diseases that occur in this region, how much it is resistant to infections and invasion by malignancies and how postsurgical and postinfection healing happen.

Keywords: cartilage, auricular, ossification, invasion, malignancies

1. Introduction

In this chapter, the practical aspects of the applied basic surgical science of the auricular cartilage are discussed on an evidence-based level according to the most recent researches in the literature.

Ear pinna (auricle) is an extremely important organ not only for the facial aesthesis but also plays a major role in hearing physiology. Both functions rely primarily on the biomechanical nature of the cartilage. Degrees of inclination and angles at its attachment to the skull determine the shape of the head and the auditory function especially the ability to localize the direction of sound.

In addition, auricular cartilage is vulnerable to many congenital and acquired diseases that require cartilage replacement or excision; this has opened the door for many advances in tissue engineering to happen. Moreover, healthy cartilage of the auricle is a plentiful source of cartilage for reconstruction of the nose, skull base and facial defects.

Consequently, comprehensive knowledge about recent advances in literature about basic science of this critical piece of cartilage is of paramount importance. In this chapter, précised, focused and between lines pieces of information will be mentioned, but old and repeated ones will not. The chapter aims to know how basic physiology, pathology, biomechanics and biochemists of the auricular cartilage can be applied to the clinical perspectives. It is not a pure clinical chapter; only related points that can be applied to the clinical practice are discussed.

Based on the abovementioned perspectives, the reader of this chapter is expected to acquire detailed knowledge about:

- Microarchitecture of the auricular cartilage and applied physiological aspect of the chondrocyte and matrix,
- Response of auricular cartilage to relapsing polychondritis as the most common autoimmune disorder affects the auricular cartilage,
- Effect of ischemia on the cartilage,
- Pathophysiology of infective chondritis and perichondritis,
- Molecular biology of invasion of the cartilage by malignancies,
- Effect of surgical intervention and trauma on the cartilage,
- Healing of auricular cartilage after surgery and trauma,
- Healing of the graft in auricle and
- Aberration of healing.

Reader also will be provided, at the end of the chapter, with an "at a glance section" that summarizes the most important advances in understandings.

2. Microstructure of auricular cartilage

The auricle is a funnel-shaped cartilaginous structure consists of a single thin plate of elastic fibrocartilage covered by skin and it is continuous with the meatus of the external auditory. It is also characterized by ridges and depressions formed by the auricular cartilage; there are five regions caused by this molding such as helix, antihelix, tragus, antitragus and concha [1].

Auricular cartilage consists of cartilage cells fill small lacunar spaces in the matrix. Young cells, called chondroblasts, are relatively small and flat and have an irregular edge with pseudopodic-type extensions lodged in the matrix. Postmitotic chondroblasts have intercellular contact and are absorbed with matrix synthesis. The chondrocytes are mature cells that grow and become spheroid with age and lose the extensions [2].

The matrix is composed chiefly of water, proteoglycans, lipids and collagens. The substance is a firm gel, positive to periodic acid-Schiff reaction, and metachromatically to toluidine blue.

The glycoproteins are a series of mucoprotein copolymers, conjoined in large lateral chains without rami, of condroitin-4-sulfate glycosaminoglycans, condroitin-6-sulfate and keratin sulfate. The proportions are modified with age, keratan sulfate increases with age [3].

The resistance to compression and the viscoelasticity are referred to their content of glycosaminoglycans, and the resistance to tension of the collagen and elastic fibers content.

2.1. Applied surgical physiology of human chondrocytes

However, porcine and bovine chondrocytes were used for many years in the tissue engineering of human auricular cartilage; human auricular chondrocytes have become the procedures of production of elastic cartilage in vitro. This has changed the future of auricular reconstruction via its marked ability to grow in tissue culture and marked ability to produce matrix of both hyaline and elastic cartilage. Human chondrocytes have the following criteria in the tissue cultures:

1. They have ability to lay down large amount of elastic cartilage under certain circumstances. Alginate-suspended aggregated chondrocytes produce matrix that contained elastin (the hallmark of the original elastic cartilage) and this amount of auricular elastic cartilage increase markedly with the alginate/collagen-containing hydrogen and enriched with k-elastin [4].
2. It can also be stimulated with insulin, dexamethasone, or growth factors such as bFGF, PDGFbb, EGF, and IGF(2.3), what is more is their ability to continue growth in the subcutaneous pocket.
3. The neo-cartilage, produced by cultured chondrocytes, does not dedifferentiate or degenerate after long cultivation time (12 weeks) and it has the same immunohistochemistry properties as the native auricular cartilage.
4. Cartilages can be created in predetermined shapes and dimensions using chondrocyte transplantation on appropriate polymer templates [4].

Ability of human auricular chondrocytes to proliferate in tissue cultures to produce auricular cartilage molds with the same histological and mechanical criteria, and the same predetermined configuration have changed surgical approaches in clinical situations such as anotia, microtia, traumatic loss and cauliflower ears. This also has replaced old-fashioned methods such as costal cartilage grafting, which was mandating the timing of surgery to be delayed until the age of 6–10 [5], and carry the risk of pneumothorax and chest wall deformities [6].

2.2. Applied surgical physiology of auricular cartilage matrix

As mentioned earlier, the biochemical composition of matrix is the factor that determines the biomechanical properties such as wear resistance, load bearing and shock absorption [7]. The mechanical properties of the auricular cartilage have not been extensively studied to date, but the Young's modulus was determined by tension calculating a modulus value for concha and tragal cartilage to be 3.4 and 2.8 MPa, respectively, but the difference was not significant [8]; however, the concha was found to demonstrate a greater Young's elastic modulus in comparison to the helix [9].

In addition, the final stress relaxation rate was similar for all five regions of the auricular cartilage, suggesting that all regions of the auricle had the ability to reach similar load equilibrium over 15 min (helix $1.78 \times 10^{-4} \pm 0.32$ MPa/s, antihelix $1.62 \times 10^{-4} \pm 0.31$ MPa/s, concha MPa/s $1.52 \times 10^{-4} \pm 0.23$ MPa/s, antitragus $1.46 \times 10^{-4} \pm 0.23$ MPa/s and tragus $1.46 \times 10^{-4} \pm 0.15$ MPa/s). The final absolute relaxation was also similar between the five regions of the auricular cartilage, demonstrating that the auricular cartilages could relax to a similar final stress level (helix 0.21 ± 0.02 MPa, antihelix 0.24 ± 0.04 MPa, concha 0.23 ± 0.04 MPa, antitragus 0.21 ± 0.03 MPa and tragus 0.23 ± 0.03 MPa) [8–10]. Therefore, surgeons can harvest any anatomical part of the auricle for reconstruction.

Such biomechanical properties are mainly due to collagen II fibers in the matrix because Dahl et al. analyzed the bimolecular composition of endogenous auricular cartilage in normal adults, pediatric patients with microtia and pediatric patients with preauricular appendages. Immunohistochemical analysis demonstrated similar levels and distribution of elastin and collagens I and X in all three groups of patients, and reduced expression of collagen II in children with microtia [11]. Collagen II is, also, the main target affected in several diseases and malignancies, which is discussed in the following paragraphs.

As mentioned earlier, reconstructive surgeon should use synthetic or tissue engineered cartilage that provides the anatomical and biomechanical properties of the human auricle to achieve good biocompatibility with the skin [12], adequate mechanical properties prevent deformation of the implant when implanted beneath the skin providing definition of the auricle shape. Also, similar mechanical properties to the surrounding tissue prevent stress at the interface [1]; mechanical mismatch can lead to micromovement between the skin and the implant when subcutaneously implanted [13], thus implant failure and extrusion.

3. Applied pathophysiology of conditions affecting auricular cartilage

3.1. Inflammatory conditions

3.1.1. Noninfective conditions

The most common disease of this type is the relapsing polychondritis (RP) that results from autoimmune reaction against collagen fibers of the cartilage. In addition, there is a condition that must be known to differentiate it from malignancies of the skin and cartilage, it is the "chondrodermatitis nodularis chronica helicis" or "Winkler's disease," which results from ischemia of the cartilage.

3.1.1.1. Relapsing polychondritis

Relapsing polychondritis (RP) is a rare multisystem autoimmune disease characterized by recurrent episodes of inflammation and progressive destruction of cartilaginous tissues, elastic cartilage of the ears and nose, hyaline cartilage of peripheral joints, vertebral fibrocartilage and tracheobronchial cartilage [14, 15]. Auricular chondritis occurs in 20% of patients at presentation and in 90% at some point during the course of the disease [15]; therefore, its applied pathophysiology will be discussed with some details in the following context. Etiology of RP is unknown but the pathogenesis is mostly due to an immunologic reaction to type II collagen in all human tissues [16, 17].

Collagen type II is the main target of the autoantibodies in RP; therefore, it is the initial step that induces the chondritis; this is approved by the fact that titers against the native type II collagen were substantially higher than titers against constituent alpha-1 (II) chains and antibodies are positive in 30% of cases [17]. This observation suggests that the antibodies were not formed after destruction of cartilage and denaturation of collagen [16]. However, not only autoantibodies against type II collagen have been detected in patients with RP but also autoantibodies against type IX and XI collagen have been found in a patient with RP [18]. In addition, recently auto antibodies against other cartilage proteins such as cartilage oligomeric matrix proteins(COMP) and matrilin-1 have been found; matrilin-1 is a cartilage matrix protein expressed at high levels in the tracheal, nasal, auricular and chondrosternal cartilage [19, 20]. Such antibodies activate both humoral and cellular immunoreaction; there are several evidences to support this [21]:

1. Damaged cartilage is infiltrated by CD4 + T-cells and plasma cells and contains immune deposits and perichondral infiltrate of lymphocytes and plasma cells with loss of basophilic staining of the cartilage matrix indicating loss of proteoglycans [22].
2. A T-cell response specific to peptides found in collagen type II (which contributes 95% of all cartilage collagen) or of matrilin-1 is found in some patients [23].
3. Over half the patients with RP carry the HLA-DR4 antigen [24, 25].
4. In one patient, oral administration of collagen type II for desensitization was apparently effective [26].

These reactions lead to severe chondritis by recruiting inflammatory cells to the cartilage, such recruitment is orchestrated by a complex cytokine network [27] such as interferon-ϒ, interleukin [IL]-2, and IL-12 [28] in addition to soluble triggering receptor (sTREM-1) expressed on myeloid cells 1, chemokine (C-C motif) ligand 4 (CCL4), vascular endothelial growth factor (VEGF) and matrix metalloproteinases-3 (MMP-3) [29].

As a result of this autoimmune reaction, many proteases are released from the inflammatory cells [21] and by chondrocytes that undergo apoptosis by the effect of MMP-3 [30], causing destruction of cartilage matrix and leading to the characteristic features of RP of the auricle.

Because collagen II is responsible for biomechanical criteria of the auricle, after repeated attacks or sometimes after a single prolonged episode, the cartilaginous structure of the ear is damaged and the pinna not only feels flabby but also may droop or even flop up and down when the patient walks [15]. Pinna also may hardened by calcifications or ossification of the connective scar tissue that replaces the cartilage. Cauliflower ear deformity occurs in about 10% of patients [22].

3.1.1.2. Chondrodermatitis nodularis chronica helicis (Winkler's disease)

Another noninfective inflammatory reaction related to the unique criteria of the cartilage in general and auricular cartilage in specific is an inflammatory lesion called chondrodermatitis nodularis chronica helicis. It is a chronic perichondritis, which is thought to be related to limited vascularity at the lateral and anterior aspect of the auricle. The skin is tightly stretched over the underlying cartilage with minimal subcutaneous tissue, which results in limited vascularity and ischemia which is thought to promote the development of this lesion [31]. Another related theory is the perichondrial vacuities which narrows the blood vessels and induce ischemia of the cartilage, leading to the clinical picture of the given disease [32]. Mostly located on the helix, this disease is characterized by a hard nodule which involves the skin and the cartilage of the ear.

Ischemia also can result from compression on the cartilage as in infection and hematoma, which are discussed in the following paragraphs.

3.1.2. Infective inflammatory conditions (perichondritis and chondritis)

Perichondritis and chondritis represent infections of the auricular perichondrium or cartilage. It is caused by blunt or penetrating trauma to the ear or by direct extension from an otitis externa. Penetrating trauma may result from various injuries, including ear piercing, assaults, bites and iatrogenic injuries. Iatrogenic infection occurs when the cartilage and soft tissues of the ear are employed as donor sites for tissue used in the repair of defects of the nose and external ear [33]. The increasingly popular piercing of the ear cartilage as opposed to the lobule may predispose to infection [34], and outbreaks have been reported and Pseudomonas is the most frequent causative organism [35]. Burn of the auricle is the most aggressive form because it makes the cartilage most vulnerable to infective chondritis due to presence of large amount of dead cartilage tissues.

Whatever the reason of chondritis, cartilage becomes intensely infiltrated with polymorphonuclear leukocytes and phagocytes, which damage the cartilage via its cytokines and inflammatory mediators [36] such as auricular cartilage, like any cartilage, lack of vascular supply; it is only supplied from the overlaying perichondrium that makes it vulnerable to ischemia and necrosis. In addition, intact perichondrium adds to the problem because it does not allow the inflammatory edema of the cartilage to expand, increasing the pressure on the cartilage which causes more necrosis and end up with cauliflower ear [37]. This pathophysiology must be applied clinically by immediate drainage of abscess and hematoma, and adequate debridement of any dead cartilage [38].

3.2. Auricular cartilage and malignancy

Cancer of the auricle accounts for around 6% of all cutaneous malignancies, out of which 50–60% are squamous cell carcinoma, 30–40% is basal cell carcinoma (BCC) and 2–6% is malignant melanomas. [39] These malignancies can invade the cartilage of the auricle via several mechanisms but the most recent mechanism rather than the direct tissue pressure effect is the role played by mediators released by the tumors.

Matrix metalloproteases (MMP) play an integral role in tumor growth and metastasis; MMPs are a family of zinc-dependent endopeptidases. They allow tumors to grow by degrading matrix barriers of the underlying cartilage and promoting angiogenesis as well as releasing active growth factors and modulating apoptosis; therefore, they are used as tumor markers malignant transformation of keratinocytes [40, 41]. Specifically, MMP-13 is associated with greater metastatic capacity and MMP-11 is linked to increased local invasiveness of SCC of the head and neck. MMP-13 (collagenase 3) preferentially degrades type II collagen found in cartilage [42]. In cSCC, MMP-13 collocates with laminin-5, which is normally founded in the basement membrane to promote keratinocyte motility to the edge of the lesion and subsequently degrades nearby tissue, allowing tumor invasion [43, 44]. Therefore, matrix of the cartilage in the auricle is an important risk factor for the squamous cell carcinoma which releases many proteolytic enzymes to facilitate invasion and spread.

4. Healing of auricular cartilage

4.1. Normal healing

Cartilage injuries can be caused by several reasons because it is liable to trauma and several surgical procedures; it is capacious source of highly resistant cartilage, it is also liable to several disfiguring congenital anomalies that necessitate plastic surgeries that require grafts to the auricle to close defects [45]. Despite the widely spread use of those grafts in auricular cartilage defects, insufficient union and loss of grafting material through absorption in the long run has regularly been reported [46].

In addition, damage associated with traumatic injuries or extensive surgical manipulation is characterized by catastrophic disruption of cartilage matrix integrity and structure, extensive chondrocyte death in the area of cartilage injury, and expansion of this "zone of injury," which is facilitated by diffusible mediators such as nitric oxide [47]. The main reason behind this is that the body does not heal isolated cartilage damage effectively because blood supply necessary for the initiation and support of the repair process is absent, a lack of sufficient stem cells to repopulate and repair the defect, and chondrocyte cell death in the surrounding cartilage which compromises tissue integrity and interferes with repair tissue integration [47]. Viable chondrocytes near the injury may proliferate, form clusters of new cells, and synthesize new matrix, but chondrocytes cannot migrate readily through cartilage tissue to the site of the injury, and the matrix components they synthesize usually are not sufficient to fill the defect [47]. To conclude, any cartilage wound healing response that does not lead to replacement of type II collagen and proteoglycan synthesis will result in tissue with abnormal morphologic and mechanical properties [48]. Unfortunately, this is the case when the basic healing process of the cartilage was studied.

Pathophysiology of healing of hyaline cartilage (auricular) can be classified according to the reason of injury:

i. Postsurgical and posttraumatic healing

General healing process of the cartilage is in the young rabbit, traumatization of cartilage perpendicular to its surface resulted within 3 days in regression and necrosis of the tissue, lining the cut end, which is neighbored by a zone of hyperactivity and increased mitosis. On the seventh day, filaments present in the matrix are arranged in bundles, which demarcate the border between the viable cartilage and the regressive zone; they are continuous with the perichondrial fibers [49], the necrotic material is invaded by macrophages and polymorphonuclear cells from the contiguous exudate. In later stages, this zone has developed into a firm layer of fibrous tissue. After 4–6 weeks, the demarcating fibers will cover the rounded stump, protecting the cartilage fragment. However, all these reactions are absent in adult rabbits making the cartilage not to heal [50].

ii. Healing of grafting to the auricle (healing at interface between graft and auricular cartilage)

Wound healing of the incision surface of the graft was similar to the reaction in the pre-existent cartilage, described earlier. Thus, the scarring occurred on both sides of the junction and therefore, the junction was in most cases, fibrous and not cartilaginous. In addition, the subcutaneous transplant site in the head and neck lead to strong inflammatory reactions and resorption of the bioartificial cartilage in contrast to orthopedic and trauma surgery where the engineered constructs or autologous chondrocytes are placed in the immunoprivileged region of joints [51].

To conclude, the end result of healing depends on the age of the patient, direction and depth of the wound as the following:

1. Large, complete-thickness defects do not heal easily; normal healing time for ear cartilage piercing is 2 months to 1 year, so intervention is a must [52].
2. Partial-thickness defects are normally repaired by deposition of fibrous scar tissue.
3. Small, full-thickness cartilage defects are replaced by fibrocartilage. The mechanism of fibrocartilaginous repair appears to be mediated by proliferation and differentiation of mesenchymal cells of the marrow [53].

Consequently, it is inevitable to find a method that enhances cartilage tissue healing or to replace the damaged cartilage as following:

1. Biologic grafts such as perichondrium have been successfully used to repair full-thickness defects, probably because the inner layer of the perichondrium, adjacent to the cartilage contains progenitor cells that can differentiate into chondroblasts [52]. However, the outer layer rapidly produces fibrous overgrowth, preventing the good cartilage-to-cartilage connection necessary to restore the mechanical function of the structure [54].
2. Tissue engineered cartilage molds can be used, as mentioned in Section 2.1.
3. Growth factors to enhance healing such as somatomedin-C have growth-promoting effect on cartilage [46]. In addition, such products that induce chondrogenesis can be produced via gene therapy. Gene therapy approaches to cartilage repair are encouraged by the ability of various gene products to enhance chondrogenesis [55]. Examples include growth factors [56] such as insulin-like growth factor-1 (IGF-1), transforming growth factor-β (TGF-β), fibroblast growth factors and various members of the BMP family, as well as transcription factors such as SOX-9 [57], certain signaling molecules such as SMADs [58], and molecules that inhibit apoptosis such as BCL-2 [59]. However, it is still difficult to administer them exogenously to sites of cartilage injury in a sustained and therapeutically useful manner.

4.2. Aberrant healing of the cartilage

Any surgical intervention, especially ear piercing, may complicate with keloid which represent one extreme of aberrant dermal wound healing that is observed only in susceptible individuals following cutaneous injury [60] with higher incidence during puberty and pregnancy, periods with hyperactivity of the pituitary gland [61]. Due to increased release of greater melanocyte stimulating hormone (MSH), keloid formation mainly occurs in parts of the body with high concentrations of melanocytes [62].

Histopathologically, keloid is included in the spectrum of fibroproliferative disorders and commonly affects the ears, it has been suggested that keloid scarring is caused by an inability to stop the wound healing process and abnormal response to inflammation by fibroblasts [63, 64]. Scar is densely populated by inflammatory cells, which release fibrogenic factors such as transforming growth factor (TGF)-β1 and -β2. This environment enhances accumulation of ECM, while its degradation is impaired (via decreased levels of TGF-β3 and matrix metalloproteinases [MMP], for example, (MMP-9) [65]. In addition, development of a Th-2 response stimulates fibrogenesis and Th-1 predominance attenuates the tissue fibrosis [66, 67].

To conclude, author summarizes this chapter in the following points, which include the latest research findings in literature about the above discussed issue.

5. At a glance

1. Both functions of the auricle, atheistic and hearing, rely primarily on the biomechanical nature of its cartilage.
2. Human auricular chondrocytes have become the procedures for the production of elastic cartilage in vitro because they lay down large amount of elastic cartilage under certain circumstances.
3. Cartilage of human chondrocytes culture is resistant to degeneration even after long time and it has the same immunohistochemistry properties the native auricular cartilage.
4. The final stress relaxation rate is similar for all five regions of the auricular cartilage; all regions of the auricle had the ability to reach similar load equilibrium over 15 min.
5. Biomechanical properties of the auricular cartilage are mainly due to collagen II fibers in the matrix which is defective in patients of congenital auricular malformations.
6. Implants must have the same mechanical properties of the cartilage otherwise mechanical mismatch can lead to micromovement between the skin and the implant thus implant failure and extrusion.
7. Collagen type II is the main target of the autoantibodies in relapsing polychondritis (RP); therefore, it is the initial step that induces the chondritis.
8. Recently, autoantibodies against cartilage oligomeric matrix proteins (COMP) and matrilin-1 have been found in patients of (RP).

9. However, not many experiments, oral administration of collagen type II for desensitization in (RP) is apparently effective.
10. Chondrodermatitis nodularis chronica helicis is proven to be partially due to ischemia of the cartilage.
11. Intact perichondrium in auricular perichondritis and hematoma adds to the problem because it does not allow the inflammatory edema or blood, respectively, to expand, increasing the pressure on the cartilage, which causes more necrosis and end up with cauliflower ear.
12. MMP-13 (collagenase 3) preferentially degrades type II collagen found in cartilage.
13. In cSCC, MMP-13 collocates with laminin-5, which is normally found in the basement membrane to promote keratinocyte motility to the edge of the lesion and subsequently degrades nearby tissue, allowing tumor invasion.
14. Chondrocytes cannot migrate readily through cartilage tissue to the site of the injury, and the matrix components they synthesize usually are not sufficient to fill the defect.
15. Healing of the cartilage depends on the age of the patient, direction and depth of the wound; large, complete-thickness defects do not heal easily and intervention is a must.
16. Subcutaneous transplant site in the head and neck lead to strong inflammatory reactions and resorption of the bioartificial cartilage in contrast to orthopedic and trauma surgery where the engineered constructs or autologous chondrocytes are placed in the immuno-privileged region of joints.
17. Keloid is one extreme of aberrant dermal wound healing that is observed only in susceptible individuals following cutaneous injury.

Author details

Mohamed Khamis Tolba Mahmoud Abdalla[1,2]*

*Address all correspondence to: mohamed_khameess@yahoo.com

1 Department of Otorhinolaryngology, Head and Neck Surgery, Alexandria University, Royal College of Surgeon of England, London, UK

2 University of South Wales, Pontypridd, UK

References

[1] Griffin MF, Premakumar Y, Seifalian AM, Szarko M, Butler PE. Biomechanical characterisation of the human auricular cartilages; implications for tissue engineering. Annals of Biomedical Engineering. 2016;**44**(12):3460-3467

[2] Quatela VC, Sherris DA, Rosier RN. The human auricular chondrocyte: Responses to growth factors. Archives of Otolaryngology – Head & Neck Surgery. 1993;**119**(1):32-37

[3] Quarto R, Campanile G, Cancedda R, Dozin B. Modulation of commitment, proliferation, and differentiation of chondrogenic cells in defined culture medium. Endocrinology. 1997;**138**(11):4966-4976

[4] Sivayoham E, Woolford TJ. Current opinion on auricular reconstruction. Current Opinion in Otolaryngology & Head and Neck Surgery. 2012;**20**(4):287-290

[5] Kawanabe Y, Nagata S. A new method of costal cartilage harvest for total auricular reconstruction: Part I. Avoidance and prevention of intraoperative and postoperative complications and problems. Plastic and Reconstructive Surgery. 2006;**117**(6):2011-2018

[6] De Chalain T, Phillips JH, Hinek A. Bioengineering of elastic cartilage with aggregated porcine and human auricular chondrocytes and hydrogels containing alginate, collagen, and kappa-elastin. Journal of Biomedical Materials Research. 1999;**44**(3):280-288

[7] Lu XL, Mow VC. Biomechanics of articular cartilage and determination of material properties. Medicine & Science in Sports & Exercise. 2008;**40**(2):193-199

[8] Zahnert T, Hüttenbrink KB, Mürbe D, Bornitz M. Experimental investigations of the use of cartilage in tympanic membrane reconstruction. Otology & Neurotology. 2000; **21**(3):322-328

[9] Griffin MF, Premakumar Y, Seifalian AM, Szarko M, Butler PE. Biomechanical characterisation of the human nasal cartilages; implications for tissue engineering. Journal of Materials Science: Materials in Medicine. 2016;**27**(1):11

[10] Kluger N, Guillot B. Body-piercing complications. Annales de Dermatologie et de Vénéréologie. 2010;**137**:153

[11] Dahl JP, Caballero M, Pappa AK, Madan G, Shockley WW, Van Aalst JA. Analysis of human auricular cartilage to guide tissue-engineered nanofiber-based chondrogenesis: Implications for microtia reconstruction. Otolaryngology – Head and Neck Surgery. 2011;**145**(6):915-923

[12] Walton RL, Beahm EK. Auricular reconstruction for microtia: Part II. Surgical techniques. Plastic and Reconstructive Surgery. 2002;**110**(1):234-252

[13] Nayyer L, Birchall M, Seifalian AM, Jell G. Design and development of nanocomposite scaffolds for auricular reconstruction. Nanomedicine: Nanotechnology, Biology and Medicine. 2014;**10**(1):235-246

[14] Gergely P, Poór G. Relapsing polychondritis. Best Practice & Research Clinical Rheumatology. 2004;**18**(5):723-738

[15] Letko E, Zafirakis P, Baltatzis S, Voudouri A, Livir-Rallatos C, Foster CS. Relapsing polychondritis: A clinical review. In: WB Saunders, editor. Seminars in Arthritis and Rheumatism. 2002 Jun;**31**(6):384-395

[16] Ebringer R, Rook G, Swana GT, Bottazzo GF, Doniach D. Autoantibodies to cartilage and type II collagen in relapsing polychondritis and other rheumatic diseases. Annals of the Rheumatic Diseases. 1981;**40**(5):473-479

[17] Foidart JM, Abe S, Martin GR, Zizic TM, Barnett EV, Lawley TJ, Katz SI. Antibodies to type II collagen in relapsing polychondritis. New England Journal of Medicine. 1978; **299**(22):1203-1207

[18] Alsalameh S, Mollenhauer J, Scheuplein F, Stöss H, Kalden JR, Burkhardt H, Burmester GR. Preferential cellular and humoral immune reactivities to native and denatured collagen types IX and XI in a patient with fatal relapsing polychondritis. The Journal of Rheumatology. 1993;**20**(8):1419-1424

[19] Buckner JH, Wu JJ, Reife RA, Terato K, Eyre DR. Autoreactivity against matrilin-1 in a patient with relapsing polychondritis. Arthritis and Rheumatism. 2000;**43**(4):939-942

[20] Hansson AS, Heinegård D, Piette JC, Burkhardt H, Holmdahl R. The occurrence of autoantibodies to matrilin 1 reflects a tissue-specific response to cartilage of the respiratory tract in patients with relapsing polychondritis. Arthritis & Rheumatology. 2001;**44**(10):2402-2412

[21] Trentham DE, Le CH. Relapsing polychondritis. Annals of Internal Medicine. 1998; **129**(2):114-122

[22] Longo L, Greco A, Rea A, Vasco VR, De Virgilio A, De Vincentiis M. Relapsing polychondritis: A clinical update. Autoimmunity Reviews. 2016;**15**(6):539-543

[23] Buckner JH, Van Landeghen M, Kwok WW, Tsarknaridis L. Identification of type II collagen peptide 261-273-specific T cell clones in a patient with relapsing polychondritis. Arthritis & Rheumatology. 2002;**46**(1):238-244

[24] Lang B, Rothenfusser A, Lanchbury JS, Rauh G, Breedveld FC, Urlacher A, Albert ED, Peter HH, Melchers I. Susceptibility to relapsing polychondritis is associated with hla-dr4. Arthritis & Rheumatology. 1993;**36**(5):660-664

[25] Zeuner M, Straub RH, Rauh G, Albert ED, Schölmerich J, Lang B. Relapsing polychondritis: Clinical and immunogenetic analysis of 62 patients. The Journal of Rheumatology. 1997;**24**(1):96-101

[26] Navarro MJ, Higgins GC, Lohr KM, Myers LK. Amelioration of relapsing polychondritis in a child treated with oral collagen. The American Journal of the Medical Sciences. 2002;**324**(2):101-103

[27] Arnaud L, Mathian A, Haroche J, Gorochov G, Amoura Z. Pathogenesis of relapsing polychondritis: A 2013 update. Autoimmunity Reviews. 2014;**13**(2):90-95

[28] Kraus VB, Stabler T, Le ET, Saltarelli M, Allen NB. Urinary type II collagen neoepitope as an outcome measure for relapsing polychondritis. Arthritis & Rheumatology. 2003;**48**(10):2942-2948

[29] Sato T, Yamano Y, Tomaru U, Shimizu Y, Ando H, Okazaki T, Nagafuchi H, Shimizu J, Ozaki S, Miyazawa T, Yudoh K. Serum level of soluble triggering receptor expressed on myeloid cells-1 as a biomarker of disease activity in relapsing polychondritis. Modern Rheumatology. 2014;**24**(1):129-136

[30] Ouchi N, Uzuki M, Kamataki A, Miura Y, Sawai T. Cartilage destruction is partly induced by the internal proteolytic enzymes and apoptotic phenomenon of chondrocytes in relapsing polychondritis. The Journal of Rheumatology. 2011;**38**(4):730-737

[31] Rickli H, Hardmeier T. Winkler's chondrodermatitis nodularis chronica helicis. Der Pathologe. 1988;**9**(1):25-29

[32] Upile T, Patel NN, Jerjes W, Singh NU, Sandison A, Michaels L. Advances in the understanding of chondrodermatitis nodularis chronica helices: The perichondrial vasculitis theory. Clinical Otolaryngology. 2009;**34**(2):147-150

[33] Kaplan AL, Cook JL. The incidences of chondritis and perichondritis associated with the surgical manipulation of auricular cartilage. Dermatologic Surgery. 2004;**30**(1):58-62

[34] Keene WE, Markum AC, Samadpour M. Outbreak of Pseudomonas aeruginosa infections caused by commercial piercing of upper ear cartilage. Journal of the American Medical Association. 2004;**291**(8):981-985

[35] Kent SE, Rokade AV, Premraj K, Butcher C. "High" ear piercing and perichondritis of the pinna. BMJ: British Medical Journal. 2001;**323**(7309):400

[36] Martin R, Yonkers AJ, Yarington CT. Perichondritis of the ear. The Laryngoscope. 1976; **86**(5):664-673

[37] Stroud MH. A simple treatment for suppurative perichondritis. The Laryngoscope. 1963;**73**(5):556-563

[38] Dowling JA, Foley FD, Moncrief JA. Chondritis in the burned ear. Plastic and Reconstructive Surgery. 1968;**42**(2):115-122

[39] Vuyk HD, Cook TD. Auricular reconstruction after Moh's surgery. A review. Faces. 1997;**5**:9-21

[40] Boisen J, Malone CH, Kelly B, Wagner RF. Cutaneous squamous cell carcinoma with invasion through ear cartilage. Case Reports in Dermatological Medicine. 2016;**16**:2016

[41] Martinez JC, Cook JL. High-risk cutaneous squamous cell carcinoma without palpable lymphadenopathy: Is there a therapeutic role for elective neck dissection? Dermatologic Surgery. 2007;**33**(4):410-420

[42] Reunanen N, Kähäri VM. Matrix metalloproteinases in cancer cell invasion. In: Madame Curie Bioscience Database. Austin, TX: Landes Bioscience; 2000. pp. 1-35

[43] Airola K, Johansson N, Kariniemi AL, Kähäri VM, Saarialho-Kere UK. Human collagenase-3 is expressed in malignant squamous epithelium of the skin. Journal of Investigative Dermatology. 1997;**109**(2):225-231

[44] Ash JE, Beck MR, Wilkes JD. Tumors of the Upper Respiratory Tract and Ear. Washington, DC: Armed Forces Institute of Pathology; 1964

[45] Sand M, Sand D, Brors D, Altmeyer P, Mann B, Bechara FG. Cutaneous lesions of the external ear. Head & Face Medicine. 2008;**4**(1):2

[46] Duncan MJ, Thomson HG, Mancer JK. Free cartilage grafts: The role of perichondrium. Plastic and Reconstructive Surgery. 1984;**73**(6):916-921

[47] Shantz JS, Marcucio R, Kim HT, Miclau III T. Chapter 4: Bone and Cartilage Healing, Section 1: General Principles: Basics, pp. 109-124

[48] Silver FH, Glasgold AI. Cartilage wound healing. An overview. Otolaryngologic Clinics of North America. 1995;**28**(5):847-864

[49] Izumi T, Scully SP, Heydemann A, Bolander ME. Transforming growth factor β1 stimulates type II collagen expression in cultured periosteum-derived cells. Journal of Bone and Mineral Research. 1992;**7**(1):115-121

[50] Zalzal GH, Cotton RT, McAdams AJ. Cartilage grafts – Present status. Head & Neck. 1986;**8**(5):363-374

[51] Rotter N, Haisch A, Bücheler M. Cartilage and bone tissue engineering for reconstructive head and neck surgery. European Archives of Oto-Rhino-Laryngology and Head & Neck. 2005;**262**(7):539-545

[52] Prudden JF, Nishihara G, Baker L. The acceleration of wound healing with cartilage. I. Surgery, Gynecology & Obstetrics. 1957;**105**(3):283-286

[53] Duynstee ML, Verwoerd-Verhoef HL, Verwoerd CD, Van Osch GJ. The dual role of perichondrium in cartilage wound healing. Plastic and Reconstructive Surgery. 2002; **110**(4):1073-1079

[54] Evans CH, Ghivizzani SC, Smith P, Shuler FD, Mi Z, Robbins PD. Using gene therapy to protect and restore cartilage. Clinical Orthopaedics and Related Research. 2000; **379**:S214-S219

[55] O'Connor WJ, Botti T, Khan SN, Lane JM. The use of growth factors in cartilage repair. Orthopedic Clinics. 2000;**31**(3):399-409

[56] Uusitalo H, Hiltunen A, Ahonen M, Gao TJ, Lefebvre V, Harley V, Kähäri VM, Vuorio E. Accelerated up-regulation of L-Sox5, Sox6, and Sox9 by BMP-2 gene transfer during murine fracture healing. Journal of Bone and Mineral Research. 2001;**16**(10):1837-1845

[57] Fujii M, Takeda K, Imamura T, Aoki H, Sampath TK, Enomoto S, Kawabata M, Kato M, Ichijo H, Miyazono K. Roles of bone morphogenetic protein type I receptors and Smad proteins in osteoblast and chondroblast differentiation. Molecular Biology of the Cell. 1999;**10**(11):3801-3813

[58] Feng L, Precht P, Balakir R, Horton WE. Evidence of a direct role for bcl-2 in the regulation of articular chondrocyte apoptosis under the conditions of serum withdrawal and retinoic acid treatment. Journal of Cellular Biochemistry. 1998;**71**(2):302-309

[59] Bran GM, Brom J, Hörmann K, Stuck BA. Auricular keloids: Combined therapy with a new pressure device. Archives of Facial Plastic Surgery. 2012;**14**(1):20-26

[60] Aköz T, Gideroğlu K, Akan M. Combination of different techniques for the treatment of earlobe keloids. Aesthetic Plastic Surgery. 2002;**26**(3):184-188

[61] Bayat A, Arscott G, Ollier WE, McGrouther DA, Ferguson MW. Keloid disease: Clinical relevance of single versus multiple site scars. British Journal of Plastic Surgery. 2005; **58**(1):28-37

[62] Gurtner GC, Werner S, Barrandon Y, Longaker MT. Wound repair and regeneration. Nature. 2008;**453**(7193):314-321

[63] Sandulache VC, Parekh A, Li-Korotky H, Dohar JE, Hebda PA. Prostaglandin E2 inhibition of keloid fibroblast migration, contraction, and transforming growth factor (TGF)-β1–induced collagen synthesis. Wound Repair and Regeneration. 2007;**15**(1):122-133

[64] Gauglitz GG, Korting HC, Pavicic T, Ruzicka T, Jeschke MG. Hypertrophic scarring and keloids: Pathomechanisms and current and emerging treatment strategies. Molecular Medicine. 2011;**17**(1-2):113

[65] Brown JJ, Bayat A. Genetic susceptibility to raised dermal scarring. British Journal of Dermatology. 2009;**161**(1):8-18

[66] Wynn TA. Fibrotic disease and the TH1/TH2 paradigm. Nature Reviews Immunology. 2004;**4**(8):583-594

[67] Doucet C, Brouty-Boyé D, Pottin-Clémenceau C, Canonica GW, Jasmin C, Azzarone B. Interleukin (IL) 4 and IL-13 act on human lung fibroblasts. Implication in asthma. Journal of Clinical Investigation. 1998;**101**(10):2129